Werkstofftechnische Berichte | Reports of Materials Science and Engineering

Reihe herausgegeben von / Edited by
F. Walther, Dortmund, Deutschland

In den Werkstofftechnischen Berichten werden Ergebnisse aus Forschungsprojekten veröffentlicht, die am Fachgebiet Werkstoffprüftechnik (WPT) der Technischen Universität Dortmund in den Bereichen Materialwissenschaft und Werkstofftechnik sowie Mess- und Prüftechnik bearbeitet wurden. Die Forschungsergebnisse bilden eine zuverlässige Datenbasis für die Konstruktion, Fertigung und Überwachung von Hochleistungsprodukten für unterschiedliche wirtschaftliche Branchen. Die Arbeiten geben Einblick in wissenschaftliche und anwendungsorientierte Fragestellungen, mit dem Ziel, strukturelle Integrität durch Werkstoffverständnis unter Berücksichtigung von Ressourceneffizienz zu gewährleisten.

Optimierte Analyse-, Auswerte- und Inspektionsverfahren werden als Entscheidungshilfe bei der Werkstoffauswahl und -charakterisierung, Qualitätskontrolle und Bauteilüberwachung sowie Schadensanalyse genutzt. Neben der Werkstoffqualifizierung und Fertigungsprozessoptimierung gewinnen Maßnahmen des Structural Health Monitorings und der Lebensdauervorhersage an Bedeutung. Bewährte Techniken der Werkstoff- und Bauteilcharakterisierung werden weiterentwickelt und ergänzt, um den hohen Ansprüchen neuentwickelter Produktionsprozesse und Werkstoffsysteme gerecht zu werden.

Reports of Materials Science and Engineering aims at the publication of results of research projects carried out at the Department of Materials Test Engineering (WPT) at TU Dortmund University in the fields of materials science and engineering as well as measurement and testing technologies. The research results contribute to a reliable database for the design, production and monitoring of high-performance products for different industries. The findings provide an insight to scientific and applied issues, targeted to achieve structural integrity based on materials understanding while considering resource efficiency.

Optimized analysis, evaluation and inspection techniques serve as decision guidance for material selection and characterization, quality control and component monitoring, and damage analysis. Apart from material qualification and production process optimization, activities concerning structural health monitoring and service life prediction are in focus. Established techniques for material and component characterization are aimed to be improved and completed, to match the high demands of novel production processes and material systems.

Weitere Bände in der Reihe http://www.springer.com/series/16102

Shafaqat Siddique

Reliability of Selective Laser Melted AlSi12 Alloy for Quasistatic and Fatigue Applications

With a Foreword by Prof. Dr. Frank Walther

 Springer Vieweg

Shafaqat Siddique
Maschinenbau/Werkstoffprüftechnik
TU Dortmund University
Dortmund, Germany

Dissertation, TU Dortmund University, 2018

ISSN 2524-4809 ISSN 2524-4817 (electronic)
Werkstofftechnische Berichte | Reports of Materials Science and Engineering
ISBN 978-3-658-23424-9 ISBN 978-3-658-23425-6 (eBook)
https://doi.org/10.1007/978-3-658-23425-6

Library of Congress Control Number: 2018956588

Springer Vieweg
© Springer Fachmedien Wiesbaden GmbH, part of Springer Nature 2019

This Springer Vieweg imprint is published by the registered company Springer Fachmedien Wiesbaden GmbH part of Springer Nature
The registered company address is: Abraham-Lincoln-Str. 46, 65189 Wiesbaden, Germany

Foreword

Research in the field of additive manufacturing addresses the aspects of additive manufacturing related to microstructure and properties, e.g. mechanical reliability. The influence of process-specific features like surface roughness, remnant porosity, residual stresses and microstructure influence the material-oriented aspects of engineering materials which do influence the property profile of resulting structures. Additionally, Department of Materials Test Engineering (WPT) at TU Dortmund University harnesses the potential of additive manufacturing techniques in developing new materials specific to additive manufacturing, which opens the door of a new era of materials performance.

The current work addresses the reliability aspects of AlSi12 alloy manufactured by selective laser melting (SLM) process. The author has investigated the thorough profile of properties starting from surface roughness, hardness, different microstructural aspects, quasistatic as well as fatigue behavior of the SLM-processed aluminum alloy. The fatigue behavior is the main focus of the work and consists of high cycle fatigue (HCF), very high cycle fatigue (VHCF), fatigue crack growth and fatigue life prediction. The influence of different in-process and post-process aspects have been investigated in terms of their processing features, resulting structural properties and corresponding mechanical behavior. The work gives a deep understanding of the mechanisms involved in the determination of the mechanical behavior of the structures manufactured by the novel process, and gives recommendations for obtaining optimal mechanical reliability.

Dortmund, July 2018   Frank Walther

TU Dortmund University
Department of Materials Test Engineering (WPT)
Baroper Str. 303, D-44227 Dortmund, Germany

Phone +49 231 755 8028
Email: frank.walther@tu-dortmund.de
Web: www.wpt-info.de

Preface

This dissertation resulted from my work as Scientific Assistant at Department of Materials Test Engineering (WPT), Technical University Dortmund. At this accomplishment, I would like to thank all the people who were involved, directly as well as indirectly, in the realization of this work.

I would like to express my gratitude to Prof. Dr.-Ing. habil. Frank Walther, head of WPT, for his dedicated and motivating supervision throughout the course of this research. His guidance as well as advanced critical, thought-provoking consultation was very important in forming the technical facets of this work. I would like to acknowledge Prof. Dr.-Ing. Claus Emmelmann, director of Fraunhofer IAPT as well as head of Institute of Laser and System Technologies (iLAS) in Technical University of Hamburg, for the intensive technical consultation as well as for providing the possibility to utilize additive manufacturing facilities. Special thanks to Prof. Dr.-Ing. Dirk Biermann for his interest in this work, as well as Priv.-Doz. Dr.-Ing. habil. Andreas Zabel for presiding over the examination commission.

Special thanks are due to Dr.-Ing. Eric Wycisk, former Scientific Assistant at iLAS, for manufacturing of the investigated material in all of its processing configurations, as well as his motivation for discussing the material aspects with regard to manufacturing conditions. I am thankful to all the colleagues, student assistants and student workers at WPT for their helpful and congenial attitude during my stay at TU Dortmund. Special thanks to my colleagues - Gerrit Frieling, Muhammad Imran and Mustafa Awd for their support in this work as well as the great camaraderie.

Finally, and most importantly, I am deeply indebted to my parents, as well as siblings for their trust and support. Not worth forgetting are the exciting dispositions by Eman, Sultan, Nahyan, Abdullah and Rida during the course of this work.

Dortmund, July 2018   Shafaqat Siddique

Abstract

Selective laser melting (SLM) is an established additive manufacturing process which can manufacture intricate parts with almost full density. The competitive advantages of the process are potential of customization as well as design optimization for light-weighting. The current state of the art recommends that utilization of the technology can now be extended from rapid prototyping to serial production for functional applications. Employment of the technology for functional applications requires not only the resulting density higher than 99%; reliability of the structures under mechanical loading is the most important parameter. To ensure reliability, influence of the processing and post-processing conditions needs to be understood. As many of the applications of SLM process are in automotive and aerospace industries, fatigue reliability of the parts plays the foremost role.

This study focuses on the investigation of process chain on the material properties and the corresponding mechanical behavior. Metallographic techniques like light and electron microscopy, non-destructive micro-computed tomography are employed to investigate the material structural properties which are then used to comprehend the mechanical behavior in quasistatic, high cycle fatigue (HCF), very high cycle fatigue (VHCF) as well as in crack propagation. Role of process-induced microstructure and defects in fatigue reliability has been discussed, and a fatigue prediction methodology has been developed based on cyclic deformation behavior, statistical analysis of defects and finite element analysis. Additionally, realizability of manufacturing hybrid structures and their influence on part reliability is also investigated.

The results indicate that the quasistatic strength of SLM-processed AlSi12 alloy is higher than that of cast alloy due to process-specific structural features, and the individual processing conditions need to be customized according to the required properties. For fatigue testing, combined load increase tests with continuously varying amplitude and constant amplitude tests help expediting the optimization procedure. SLM can be used for generation of parts with localized properties. Combination of conventional and additive manufacturing can be successfully carried out with appropriate post-processing. The developed fatigue prediction methodology can be applied for SLM structures to help reduce testing effort. Finally, the essential aspects, still need to be investigated to fully exploit the potential of SLM technology, are highlighted.

Kurzfassung

Das selektive Laserstrahlschmelzen (engl. selective laser melting – SLM) ist ein bewährtes additives Fertigungsverfahren, das die Herstellung komplexer Bauteile mit geringer Porosität ermöglicht. Die Vorteile des Verfahrens liegen in der Anpassungsfähigkeit und in den Möglichkeiten der Designoptimierung für Leichtbauanwendungen. Der aktuelle Stand der Technik legt nahe, dass das Haupteinsatzgebiet vom Prototypenbau in die Serienfertigung für Funktionsbauteile übergeht. Für beide Einsatzgebiete ist eine Dichte von mehr als 99% sowie ausreichende Kenntnis über die Zuverlässigkeit unter mechanischer Last Grundvoraussetzung für den industriellen Einsatz der Bauteile. In diesem Zusammenhang ist ein grundlegendes Verständnis der Einflussfaktoren sowohl aus dem Fertigungs- wie auch dem Nachbehandlungsprozess obligatorisch. Da das Hauptanwendungsgebiet für das selektive Laserstrahlschmelzen in der Automobil- und Luftfahrtindustrie liegt, ist das Ermüdungsverhalten von zentraler Bedeutung.

Im Fokus der Arbeit steht die Untersuchung des Einflusses der Prozesskette auf die Werkstoffeigenschaften und das dadurch bestimmte mechanische Verhalten. Um die strukturellen Eigenschaften zu untersuchen, werden metallografische Methoden, wie die Licht- und Elektronen- mikroskopie sowie die zerstörungsfreie Mikro-Computertomografie, eingesetzt. Diese Ergebnisse stellen die Grundlage für die Korrelation mit dem mechanischen Verhalten bei quasistatischer und zyklischer Beanspruchung mit hohen und sehr hohen Lastspielzahlen (HCF-VHCF) sowie bei der Rissausbreitung dar. Der Einfluss der prozessbedingten Mikrostruktur und Defekte auf das Ermüdungsverhalten wird diskutiert. Zudem wird ein Modell zur Vorhersage der Ermüdung auf der Grundlage des zyklischen Verformungsverhaltens, der statistischen Auswertung von Defekten und der Finite-Elemente-Methode entwickelt. Zusätzlich wird die Realisierbarkeit von Hybridstrukturen und der Einfluss dieser Strukturen auf die Bauteilzuverlässigkeit untersucht.

Die Ergebnisse zeigen, dass die quasistatische Festigkeit einer mittels SLM gefertigten AlSi12-Legierung im Vergleich zu einer Gusslegierung aufgrund der prozessbedingten strukturellen Eigenschaften höher ist und die individuellen Prozessbedingungen an die Werkstoffanforderungen anzupassen sind. Im Rahmen der Ermüdungsversuche wurden sowohl Laststeigerungsversuche mit kontinuierlich ansteigender Spannungsamplitude als auch Einstufenversuche durchgeführt. Durch die Kombination dieser beiden Versuchsarten besteht die Möglichkeit den Optimierungsprozess zu beschleunigen. Insbesondere kann das SLM-Verfahren dafür verwendet werden, die Bauteileigenschaften lokal zu beeinflussen. Die Kombination von konventioneller und additiver Fertigung ist mit entsprechender Nachbehandlung erfolgreich realisierbar. Das entwickelte Vorhersagemodell zur Beschreibung des Ermüdungsverhaltens kann auf SLM-Bauteile angewandt werden, um den Versuchsaufwand zu reduzieren. Abschließend werden die wesentlichen Aspekte hervorgehoben, um das Potenzial des SLM-Verfahrens vollständig auszunutzen.

Table of contents

List of abbreviations

Abbreviation	Description
2D	Two-dimensional
3D	Three-dimensional
AM	Additive manufacturing
ANOVA	Analysis of variance
BPH	Base plate heating
CAD	Computer aided design
CAT	Constant amplitude test
CT	Computed tomography
DED	Directed energy deposition
DMLS	Direct metal laser sintering
EBF	Extrusion-based fusion
ED	Energy density
EDX(S)	Energy dispersive X-ray (spectroscopy)
FDM	Fused deposition modeling
FEM	Finite element modeling
HAZ	Heat affected zone
HCF	High cycle fatigue
HIP	Hot isostatic pressing
HV	Hardness Vickers
LAM	Laser additive manufacturing
LC	Laser cusing
LCF	Low cycle fatigue
LIT	Load increase test

MANOVA	Multivariate analysis of variance
MMC	Metal matrix composite
NDT	Non-destructive testing
OPS	Oxide polishing suspension
PBF	Powder bed fusion
PEEQ	Equivalent plastic strain
PH	Precipitation hardening
RM	Rapid manufacturing
RP	Rapid prototyping
SEM	Scanning electron microscopy
SH	Servo hydraulic
SLM	Selective laser melting
SLS	Selective laser sintering
SP	Shot-peening
SR	Stress-relief
STL	Standard tessellation language
USF	Ultrasonic fatigue
UTS	Ultimate tensile strength
VHCF	Very high cycle fatigue
XCT	X-ray computed tomography

List of symbols

Latin symbols

Symbol	Description	Unit
a	Crack length	mm
b'	Material strength exponent	-
c	Fatigue ductility exponent	-
C	Material constant of Paris equation	-
d	Hatch distance	mm
da/dN	Crack propagation rate	m/cycle
e	Expected error	-
E	Young's modulus	GPa
E_v	Volume energy density	J/mm^3
f	Frequency	Hz
F	Force	N
I	Intensity of photons	W/m^2
I'_n	Non-dimensional parameter	-
K	Stress intensity factor	MPa·$\sqrt{m}$
K_{IC}	Fracture toughness	MPa·$\sqrt{m}$
K_t	Stress concentration factor	-
m	Slope of Paris line	-
n'	Cyclic hardening exponent	-
N_f	Number of cycles to failure	-
P	Power	W
P_{max}	Maximum hydrostatic pressure	MPa
R	Stress ratio	-

R_b	Build rate	mm³/s
t	Time	s
t	Layer thickness	mm
T	Temperature	°C
u	Displacement	m
v	Scan speed	mm/s
x	Distance	m
Y	Geometry factor	-
z	Deviation at a certain confidence level	-

Greek symbols

Symbol	Description	Unit
α	Weibull scale parameter	-
β	Weibull shape parameter	-
ΔK	Stress intensity factor range	MPa·√m
ΔK_c	Critical stress intensity factor range	MPa·√m
ΔK_{eff}	Effective stress intensity factor range	MPa·√m
ΔK_{th}	Threshold stress intensity factor range	MPa·√m
$\varepsilon_{a,p}$	Plastic strain amplitude	-
ε_t	Total strain	-
ε'_f	Fatigue ductility coefficient	-
μ	Linear attenuation coefficient	-
ρ	Density	kg/m³
σ_{max}	Maximum stress	MPa
σ_{min}	Minimum stress	MPa
σ_m	Mean stress	MPa
σ_a	Stress amplitude	MPa

σ_{UTS}	Ultimate tensile strength	MPa
$\sigma_{0.2\%}$	Proof stress	MPa
σ_{ym}	Microscopic yield stress	MPa
σ_{DV}	Dang-Van equivalent stress	MPa
σ_f'	Fatigue strength coefficient	-
γ	Pore characteristic	-
λ	Thermal conductivity	W/m·K
c'	Material strength exponent	-
$\Delta\varepsilon_p$	Plastic strain range	-
T_{max}	Maximum shear stress	MPa
φ	Sensitivity factor	-

1   Introduction

Selective laser melting process has the capability of producing intricate designs which, otherwise, would be quite difficult, time-consuming and costly to be manufactured. It also offers competitive advantage to decrease the product development cycle not only for ensuring geometrical integrity but for making prototypes for functional testing. Owing to the advancements in powder quality, laser technology and SLM machine systems, selective laser melting (SLM) cannot be now limited to prototyping only, but be extended to serial production. It is establishing itself as an appropriate manufacturing technology for several industries – from automotive to aerospace and to medical applications. For all the moving parts, and especially flying parts, manufacturing by additive processing can reduce the moving weight due to its potential of producing the components of virtually any geometrical complexity, encouraging weight-reduction by topology optimization and shifting the design philosophy from 'design for manufacturing' to 'design for performance'. Aerospace industry expects sufficient reduction of weight of flying components by using additive manufacturing; Airbus for example has projected 30% weight-reduction by utilization of additive manufacturing. Aluminum and its alloys are of interest to many industries due to its lower density.

For the SLM parts to be applied as functional components, their performance needs to be carefully investigated so that the functional reliability can be ensured. The performance metrics of the SLM parts should at least be at par with those of conventionally-manufactured, currently employed parts. These metrics include mechanical properties under quasistatic and fatigue loading, which depend on the material parameters like material integrity, composition as well as microstructure. Parts produced by additive manufacturing have their quasistatic performance comparable to that of conventionally-manufactured alloys, even better for some alloys; however, the potential application areas of SLM – automotive, aerospace and medical industries – suggest that the parts are loaded not only under static or quasistatic loading, but fatigue as well. Fatigue failure has been observed for many automotive, locomotive as well as aerospace accidents, and therefore fatigue reliability is an important aspect to be ensured before the SLM parts can be employed to functional components.

Current state of the art regarding feasibility of SLM parts in functional applications limits the demonstrability for their employment in applications requiring mechanical reliability. Many available studies focus on the processibility of various alloys with corresponding microstructure and are mostly limited to determination of quasistatic properties. Literature investigating the fatigue

© Springer Fachmedien Wiesbaden GmbH, part of Springer Nature 2019
S. Siddique, *Reliability of Selective Laser Melted AlSi12 Alloy for Quasistatic and Fatigue Applications*, Werkstofftechnische Berichte │ Reports of Materials Science and Engineering, https://doi.org/10.1007/978-3-658-23425-6_1

performance of SLM parts is rather scarce. In the last few years, some studies have reported the fatigue behavior of SLM parts of titanium, aluminum and steel alloys. These studies comprise of islands of scientific information exploring one phenomenon or the other. A comprehensive study investigating the influence of processing parameters on the part parameters and, in turn, on the mechanical properties is required to be carried out.

The current study, therefore, is an attempt to investigate the influence of some of the processing and post-processing parameters on the material structure and the corresponding material properties. The objective is to investigate the underlying phenomena which occur in this relatively novel manufacturing process where each parameter may influence the structural parameters in a peculiar way which will result in the corresponding properties. The investigated property profile ranges from quasistatic tensile behavior to high cycle fatigue to very high cycle fatigue (in gigacycle range) as well as crack propagation behavior. The mechanical behavior is related to processing parameters by investigating hardness, microstructural parameters, process-induced defects as well as residual stresses.

To achieve these goals, this dissertation assumes a step-by-step approach. As it is an interdisciplinary topic, the first few sections in chapter 2 are meant for developing the understanding and relationships between these specific fields. Section 2.1 introduces to the principles and types of additive manufacturing techniques in general and SLM process in particular with the elaboration of the SLM parameters of importance. It then gives an overview of the property profile of the SLM-manufactured materials including microstructure, relative density and residual stresses. Section 2.2 gives description of the material failure under cyclic loading i.e. fatigue failure, illustrating the tools of fatigue characterization, introducing to different ranges of fatigue regarding number of cycles to failure as well as the fatigue mechanisms in different phases of fatigue life i.e. crack initiation and crack propagation. It also gives a synoptic view of material parameters which are responsible for the corresponding fatigue behavior, such that these parameters can be studied in relation to the property profile of SLM parts discussed in section 2.1.

Section 2.3 compiles the existing literature regarding the effect of SLM process parameters on different properties succinctly. It arranges the influence of the already investigated parameters on part properties, quasistatic properties, fatigue behavior as well as crack propagation behavior. It then analyzes the state of the art and, based on the existing knowledge voids, identifies the objectives of this dissertation. Before starting description of experimental details, section 2.4 elaborates the characteristics of aluminum-silicon alloys so that a better under-standing of the loading conditions and processing can be developed where the investigated material is taken into account.

Experimental procedures to achieve the objectives are described in chapter 3. Design of experiments to analyze the effects of different parameters is given there together with the techniques used for metallography, microscopy, quasistatic and fatigue loading as well as analysis procedures. Chapter 4 portrays the results of experimental characterization including powder characteristics, microstructure, chemical composition, physical properties and residual stress profile. Quasistatic and fatigue behavior of the investigated specimens is discussed in section 5.1 where the influence of different processing parameters on quasistatic properties, high cycle fatigue, very high cycle fatigue and crack propagation behavior is reported.

Another feature investigated in this dissertation is the possibility of manufacturing hybrid structures i.e. by combination of conventional and additive manufacturing processes. Section 5.2 presents a strategy for hybrid specimens to determine their potential for quasistatic and fatigue applications. Section 5.3 discusses the results of fatigue prediction methodologies where fracture mechanics-based approach and plasticity-based approaches have been employed for fatigue life prediction of pure AlSi12 alloy as well as hybrid specimens. Predicted values are compared with the experimental ones and corresponding interpretations are made.

Chapter 6 summarizes the investigations where relationships between process parameters, structural properties and mechanical behavior are abridged, and the dissertation concludes in chapter 7 with an outlook for further research work to be carried out in this field.

2 State of the art

2.1 Additive manufacturing (AM)

Additive manufacturing (AM) is a free-form manufacturing technique which takes three-dimensional computer aided design (3D-CAD) model as input, adding raw material gradually, usually in a computer-controlled layer-wise manner, to build the part by consolidating the raw material applying an energy source. Additive manufacturing processes are in place for a number of decades; however, it is only possible since a number of years that metallic parts with relative densities higher than 99% can be manufactured, posing the opportunity to shift the potential of the technique from prototyping to serial production. This chapter gives a brief overview of the existing additive manufacturing processes regarding source of energy, type of raw material, processing capabilities and issues.

2.1.1 Spectrum of AM processes

As a broad classification, additive manufacturing processes can be segregated based on the material type – polymer and metal additive manufacturing [1,2]. Polymer-based additive manufacturing, generally termed as three-dimensional printing (3D printing), is usually used for rapid prototyping and finds less functional applications where they are subjected to cyclic loading [3-5]. As this study focuses on the mechanical performance of metallic parts, only metal-based processes will be introduced. Polymer-based additive manufacturing exists from a number of decades and includes processes like stereolithography, laminated object manufacturing, fused deposition modeling, selective laser sintering etc., and their details can be found abundantly in literature [1,6,7]. The metal additive manufacturing processes can be classified, according to Gibson [1], as follows:

A Powder bed fusion (PBF)

Powder bed fusion (PBF) processes are usually the variants of the process schematized in Fig. 2.1. Powder material is laid by the leveling instrument onto the platform in the form of a thin layer which is exposed to the source of energy, laser or electron energy, melting the powder material selectively. After that a new layer is introduced and the process repeats itself to build the complete part. Laser sintering or melting is the most common of the metal additive manufacturing processes. Usually an inert gas is used to avoid oxidation. The process has different variants: direct metal laser sintering (DMLS), selective laser sintering (SLS),

© Springer Fachmedien Wiesbaden GmbH, part of Springer Nature 2019
S. Siddique, *Reliability of Selective Laser Melted AlSi12 Alloy for Quasistatic and Fatigue Applications*, Werkstofftechnische Berichte | Reports of Materials Science and Engineering, https://doi.org/10.1007/978-3-658-23425-6_2

selective laser melting (SLM), laser cusing (LC) etc. These variants are mostly variations of the basic approach to avoid patent issues [8,9]. Different fusion mechanisms exist in the PBF processes and are described below briefly.

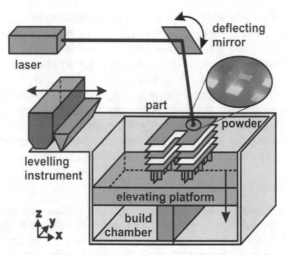

Fig. 2.1:   Generalized schematic of powder bed fusion-based additive manufacturing [10]

Solid state sintering

Solid state sintering is the term used when the powder particles are not melted completely, but diffused together in the solid state at higher temperatures. The sintering temperature is usually significantly less than melting point, therefore less fusion energy is required which results in non-dense parts having inherent porous internal volume. The porosity can be minimized by using smaller size powder particles and longer sintering time. Longer sintering time, however, is not beneficial for economic reasons causing lower build rate [11].

Chemical-induced binding

Chemical-based sintering makes use of another medium which reacts to bind the particles of the major powder material together. The binding material can be powder particles of another material, or some gases can be utilized to induce the binding. The examples include production of SiC parts by heating them to very high temperatures to cause disintegration of Si and C elements. Free silicon reacts with atmosphere to result in SiO_2 which binds the particles together. Major issue with chemical-induced binding is part porosity, requiring the need of post-processing for applications requiring full density [11,12].

Liquid phase sintering

It is a sectional melting procedure, where a portion of the constituent particles are melted which is used to bind the remaining particles without sintering them. There are many variations of this technique. One of these includes different binder and structural grains. Usually metals are used as the binder material with much smaller size compared to structural material which results in economical melting of these small particles which bind the structural parts [13]. In another alternative, composite grains are used where the powder grains contain the mixture of structural grains and the binder. Another possibility is to coat the structural grains with the binder material which ensures that the binder material at the periphery absorbs the energy preferentially, resulting in a relatively effective binding [14,15].

Full melting

Full melting is applied to the applications where fully dense parts are required. The energy source melts the through thickness of the powder material and this layer is solidified until the next layer is melted. Due to rapid melting and solidification mechanism associated, they usually possess distinct microstructure resulting in high strengths. The relative density achieved by this mechanism, selective laser melting for instance, has reached more than 99%. To achieve the dense parts, high energy input and exposure time is required [16–18].

B   *Extrusion-based fusion (EBF)*

Extrusion-based additive manufacturing uses material extruded from a nozzle head to be placed onto the platform or over the existing material layer to form a part. The material coming out of the nozzle head is in semi-molten form and solidifies over the already extruded part being an integral part of the whole structure. Temperature of the material is used to control the flow of the material. The most common extrusion-based technique is developed by *Stratasys* as fused deposition modeling (FDM). FDM has issues in accuracy and material density. As the layer thickness gets bigger, the relative density of the material decreases. Additionally, it has a limitation in forming sharp corners, as the available nozzles are only circular [19,20].

C   *Directed energy deposition (DED)*

A part is produced here by melting the material during deposition process. This is contrary to the powder fusion processes where the energy source is used to melt the already deposited powder material on the platform. In the directed energy deposition, the feedstock material is melted by laser or electron beam at the same time as it is being deposited, which implies that the melt pool is traveling. It can therefore be used to manufacture complex structures. The microstructure of these parts is usually similar to those of powder bed fusion. The process can be used to control the material or element percentage at any specific location / layer(s) by adjusting the process parameters and / or by applying multi-nozzle systems, so functionally-graded structures can be manufactured with this strategy. The process has usually a lower resolution and higher surface roughness, and denser support structures are required for complex parts [21,22].

2.1.2 Selective laser melting (SLM) process

It is a type of powder-bed process suitable for customized or complex geometries which are difficult or expensive to be manufactured by conventional processes. It overcomes the constraints of conventional processes such as tooling and physical access to surfaces for machining. These constraints previously restricted the freedom-of-design. Most of the applications of SLM are in light-weight parts for aerospace or individual and complex parts for medical applications. Using this technique, parts can be produced virtually of any design. The process is a revolution in rapid prototyping (RP) and rapid manufacturing (RM), as the part can be available in tangible form within a few hours after design completion, reducing the product development time. Moreover, fully functional parts can be manufactured [23]. *Fockele* and *Schwarze* with co-operation of *Fraunhofer Institute of Laser Technology (ILT)* developed the first SLM system [24].

Fig. 2.2 shows the typical processing sequence of the SLM process. 3D-CAD model in STL (standard tessellation language) format is given as input to the SLM system, which is sliced into small layers according to the defined layer thickness such that they can be treated as two dimensional layers. The scanner scans the corresponding cross-section of the model. The corresponding powder material is fused by the absorbed thermal energy. The part is produced on a building plate based on a movable platform. After melting a layer, the substrate plate moves down one layer thickness in z-direction and the process is repeated for the next layer. The process continues till the entire part is manufactured [24–26].

The STL model can be designed by any CAD software or can be generated from an existing component making use of reverse engineering. Support structures must be added to the design for facilitating part removal from the base plate. They not only support the component placement when the surrounding material is untied powder, but also prevent bulging [27]. A base plate is used as the building platform which is usually of the same / similar alloy which is to be manufactured. The difference in materials of the powder and base plate can cause thermal distortion. Base material can be continuously heated to a specific temperature to reduce thermal gradients. The powder material is melted by a single or multiple sources of laser which should impart an energy which is sufficient for melting the powder layer. A reservoir of powder material is an integral part of the SLM machine system which spreads the defined thickness of powder material using a leveling instrument. The powder material is part of a closed-loop system such that the un-melted powder can be re-used together with the new material [1,25,26].

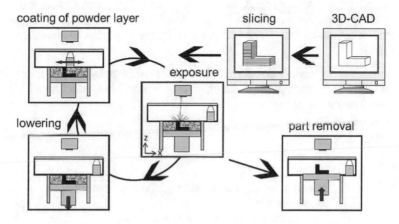

Fig. 2.2: Manufacturing sequence in selective laser melting (SLM) process [25]

Parameters of the SLM process

The performance of SLM process in terms of part quality depends on many parameters. These process parameters can directly or indirectly influence the part parameters like density, roughness, residual stresses, hardness, quasistatic and fatigue properties. The most important part property in such technique is porosity. If some powder particles remain un-melted, they can be a source of stress concentration during mechanical loading, acting as onset of plastic deformation resulting in early failures [28]. To ensure consistent and reliable part properties for reliable applications, process parameters with optimal part properties need to be ensured with repeatability and reproducibility. The dominant process features influencing part properties are laser power, scan speed, hatch distance, layer thickness and scanning strategy. The influence of these parameters has been a topic discussed in many of the studies [29–31]. The most important parameters are discussed below.

Laser

Energy of the laser is absorbed by the powder material, which melts due to absorbed energy. Choice of the used laser may depend on the material being manufactured, as the wavelength determines the absorptivity [32]. Different types of lasers are used in SLM systems with different laser powers (P). Initially CO_2 lasers, wavelength of 10.6 µm, were used with a power of up to 200 W [14]. Later, solid state lasers were developed. Currently the lasers used are fiber lasers with a wavelength of 1.06 µm having finer spot size and can achieve a higher scan resolution. With a power of 300 to 1000 W, these high power lasers with high absorptivity at smaller wavelength have ensured complete melting of the powder material as compared to partial melting with CO_2 lasers in selective laser sintering (SLS) [33].

How much volume energy density (E_v) is imparted to the powder material depends on laser power (P) scan speed (v), hatch distance (d) and layer thickness (t) which are related in the following way:

$$E_v = \frac{P}{v \cdot d \cdot t} \tag{2.1}$$

where:

E_v [J/mm³]:   Energy density imparted to the powder material
P [W]:   Laser power
v [mm/s]:   Scan speed at which laser beam travels
d [mm]:   Hatch distance, the lateral distance between two consecutive laser tracks
t [mm]:   Thickness of one layer

An important parameter, based on these factors, is the build rate, R_b [mm³/s] which is a measure of the productivity of the process as portrayed in eq. 2.2.

$$R_b = v \cdot d \cdot t \tag{2.2}$$

These factors, combined in the form of energy density, together with scanning strategy, determine the melting structure. Scanning strategy determines the orientation of the part.

A general overview of the parameters of scanning is shown in Fig. 2.3a. If all the layers are scanned in the same direction (layer n-1 or layer n in Fig. 2.3b), the resulting part will exhibit directional microstructure, and therefore anisotropic properties. However, for favoring uniaxial or isotropic microstructure, a mixed scanning strategy, where each layer scanning is rotated with reference to the previous layer, can be employed. Usually a mixed scanning strategy is employed where different scanning strategies are used for contours and core to ensure the complete melting at contours. Different machine systems with available chamber capacity and process parameters are commercially available [8,34,35].

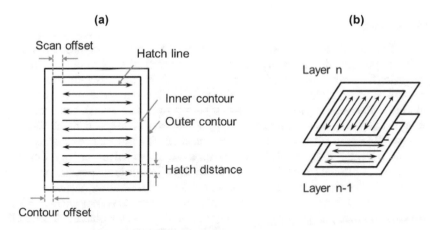

Fig. 2.3: Schematic explaining scanning parameters (a); rotation of scanning orientation in each layer (b)

2.1.3 Property profile of SLM-manufactured parts

The SLM process has its specific process parameters and the resulting part properties are a direct result of these specific process features. The basic characteristics of SLM parts, as obtained by specific process parameters, are discussed below.

Dimensional accuracy

One of the main advantages of SLM process is manufacturing of parts having characteristics of complex geometry, thin walls and other geometrically intricate features. An example from medical applications is orthodontic brace with customized brackets. Every individualized bracket contains its own curved bonding which is compatible with the corresponding tooth [36]. Therefore, the foremost quality parameter is the ability of the process to manufacture these small features with dimensional stability. The parameters which may affect the geometrical accuracy are laser power, thickness of one layer, the scanning speed as well as the build orientation [36–38]. Powder material, its size and the expansion ratio, also influences the accuracy [39,40]. Additionally, the conversion of CAD data to STL format determines the part surface by using triangular facets. The underlying resolution of this data conversion also influences the geometrical accuracy [26]. The current dimensional accuracy of the commercial systems lies between 100-150 µm [36,41].

Relative density

For the functional components, relative density of the parts as compared to their maximum possible density is important. This is specifically relevant for dynamic applications where any remnant porosity can affect the strength drastically. Obtained density of the SLM parts depends on the energy density (eq. 2.1) imparted to the material and is a function of laser power, scan speed, hatch distance and layer thickness. Lower laser power, higher values of scan speed, hatch distance and layer thickness all result in improved productivity but reduced part quality carrying bonding defects [42]. However, even at constant energy density, differences in remnant porosity are observed at different laser powers and scan speeds [43]. The porosity in SLM process can be due to balling phenomenon or stress-caused cracking. The achieved density by SLM process now exceeds 99%. The small remnant porosity has only a very minor effect on the quasistatic strength of the materials, and their tensile strength is usually comparable or higher than those of conventional alloys; however, the remnant porosity may create stress concentration in the component and cause early failures. Low thermal gradients can further improve the relative density by avoiding stress cracking and degassing [18].

Microstructure

SLM process offers a unique microstructure representative of its layered manufacturing and high cooling rates. The laser processing parameters which influence imparted energy density and processing speed determine the type of microstructure [42-44]. The very high cooling rates in SLM process (10^5-10^6 K/s) limit the growth of grains which results in needle-like or dendritic structure in most of the metals [45–47]. The size of these needles or dendrites is as low as a few hundred nanometers [10,49,50]. The microstructure can be varied in-process by allowing the grains to grow when the thermal gradients are lowered. It can be obtained by pre-heating the base plate [51,52], or by post-process heat treatments [47,53]. Laser power significantly influences the achieved micro-structure by changing the cooling rate [54].

Surface quality

Surface quality of the parts is important for the application of parts in cyclic applications. For fatigue performance of any material, surface roughness influences the crack initiation at the surface where rough surfaces can act as micro-notches. Average roughness R_a of the materials manufactured by SLM is found between 5-15 µm, and the maximum roughness R_z in the range of 70-100 µm [55]. Re-melting of the contours can significantly improve surface quality up to average roughness R_a of 2 µm [56–58]; however, adding to the production cost.

Residual stresses

Residual stresses are usually developed in SLM parts due to high thermal gradients in the process, localized heating as well as volume compaction of the molten material as compared to powder material [27,42]. Large residual stresses have been reported in SLM parts. These residual stresses may cause distortion in the part and result in cracking [59]. The stresses are not consistent within a part; they tend to increase with the height of the substrate. If thermal gradients are decreased, e.g. by heating the substrate plate, the residual stresses are significantly reduced [60]. Tensile residual stresses may increase the amplitude of applied stresses and reduce the effective strength of the material. However, compressive residual stresses may have an opposite effect. In the SLM process, the process gives a potential to control the process parameters in a way to induce residual stresses at the desired location resulting in graded properties.

Hardness

SLM-manufactured parts can have varying levels of hardness. In the presence of porosity, hardness is decreased as the pores collapse when load is applied [61]. Hardness of SLM parts manufactured with optimal process parameters is usually comparable or better than that of conventionally-manufactured alloys, with higher hardness as a consequence of fine microstructures, additional to other features like type of microstructure, phase orientation, for instance [54]. The hardness can be varied by controlling the cooling rate in-process or by post-process heat treatment.

2.2   Fatigue behavior of metallic materials

Fatigue failure is the failure of materials or structures under cyclic loading. Although materials can withstand a single load cycle below their quasistatic strength; if this load cycle is repeated many times, it can initiate failure of the material below its yield strength.

2.2.1   Cyclic deformation behavior and Woehler curve

Fatigue process can change the stress-strain behavior of materials. Though dominant loading in fatigue process is elastic, cyclic plastic deformation occurs in microscopic range. Changes in material properties due to cyclic loading can be realized by comparing quasistatic and cyclic deformation curves. Cyclic hardening of the material will increase the stress level at a specific strain as compared to strain in quasistatic loading, and vice versa. The level of cyclic hardening or softening depends on the microstructural changes, specifically movement of dislocations and their interaction with each other and with other lattice defects. For soft materials, usually more dislocations occur under cyclic loading increasing the density which, by decreasing the motion of dislocations, causes the material to harden. To the contrary, less number of dislocations favor their movement, resulting in cyclic softening of the material [62,63]. When cyclic deformation behavior is expressed by changes in plastic strain amplitude ($\varepsilon_{a,p}$), cyclic softening will be represented by increase in $\varepsilon_{a,p}$ (Fig. 2.4a) and cyclic hardening by decrease in $\varepsilon_{a,p}$ (Fig. 2.4b). A mixture of the two mechanisms i.e. softening and hardening (Fig. 2.4c) is usual for many materials.

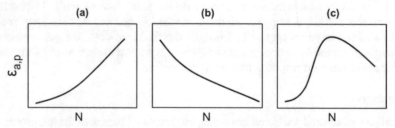

Fig. 2.4:   Exemplary cyclic deformation behavior in terms of plastic strain amplitude ($\varepsilon_{a,p}$) as a function of number of cycles (N) showing: cyclic softening (a), cyclic hardening (b), cyclic softening followed by cyclic hardening (c), adapted from [63]

Cyclic softening and hardening are responsible for the fatigue life of a material. Mostly quasistatic deformation properties are used for design of structures even in the cyclic applications, as these are the easiest to be obtained. Keeping in view the time-extensive fatigue tests, attempts have been made to find correlations between cyclic strength and quasistatic properties e.g. research by Smith et al. [64] has predicted hardening and softening behavior by quasistatic tensile strength and yield strength. Landgraf et al. [65] have also proposed cyclic deformation based on strain hardening exponent. Basan et al. [66] have developed correlations between cyclic deformation behavior and hardness of steel materials. However, all these methods are material-dependent. There is a need to relate the cyclic deformation behavior with the fatigue life of materials.

Fatigue life of a material or a component is represented classically by so-called Woehler curve which is a graph between applied stress amplitude and the resulting number of cycles to failure. The graph can also be represented in the form of Coffin-Manson plot where plastic strain amplitude is plotted against number of cycles to failure as shown in Fig. 2.5 proposed by Mughrabi [67] from low cycle fatigue to very high cycle fatigue.

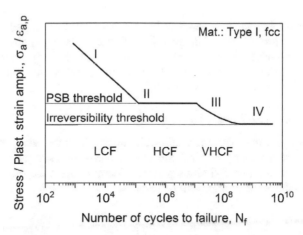

Fig. 2.5:   Representation of fatigue life ranges from conventional low cycle fatigue (LCF) to state of the art very high cycle fatigue (VHCF) in terms of number of cycles to failure (N_f) vs. stress amplitude (σ_a) or plastic strain amplitude ($\varepsilon_{a,p}$), adapted from [67]

The general behavior of fatigue life is that it decreases with increasing stress or strain amplitude. Conventionally, the fatigue ranges were classified as low cycle fatigue (LCF) and high cycle fatigue (HCF) and the fatigue tests were carried out usually until 10^7 cycles for the metals without endurance limit and $2 \cdot 10^6$ cycles for metals with endurance limit. Ranges I and II represent the conventional Woehler curve where range I follows Coffin-Manson law. Horizontal asymptote in range II was considered as "fatigue limit" and the stresses below that level were considered to result in unlimited fatigue lives without failure. This range II may not be considered as the real fatigue limit but can be considered as a threshold of persistent slip bands (PSB) and can be called as plastic strain fatigue limit. The availability of high frequency testing systems has now made it possible to test the fatigue lives in the gigacycle range, the so-called very high cycle fatigue (VHCF) range. Research in the VHCF range has questioned the existence of fatigue limit and the current research suggests that the material fails even at stresses below the conventional fatigue limit with fatigue life increasing with decreasing stress amplitude.

Range III may be termed as a transition range from HCF to VHCF which is explained by accumulation of very small localized stress raisers until a so-called "embryonic" PSB is formed which may become pronounced enough that a fatigue crack is initiated. In this range, therefore, fatigue life is dominated by the life in crack initiation phase. A real fatigue limit can be expected only in range IV which is the irreversibility threshold of the VHCF limit. These mechanisms are valid for type I materials i.e. face centered cubic (fcc) materials without any metallic inclusions and internal material defects. Type II materials, e.g. high-strength steels as well as multi-phase titanium alloys undergo sub-surface fatigue crack initiation from material defects and the defined ranges are more vague due to expected scatter related to material defects [68–70].

2.2.2   Parameters affecting fatigue behavior

A number of factors like material yield strength, roughness, mean stress, temperature, environment, residual stresses, material defects etc. influence the fatigue behavior of a material or a structure. However, here the factors which are influenced by the SLM process will be discussed.

Relative density

Parts short of their full theoretical density would mean that there exists some porosity in the part. On micro-level, porosity affects stress distribution and act as a favorable site for crack nucleation. Although the nucleation usually starts at the surface; in case of pores, it can also start sub-surface if there is a pore present slightly sub-surface. Crack initiation from internal pores can be caused by several factors including size of the pore, location of the pore as well as the applied loading. Small size pores existing far below the surface offer less possibility of crack nucleation. Though the parts manufactured by SLM have a relative density higher than 99%, the remnant porosity of less than 1% may significantly influence the fatigue strength. The pores of size more than 100 μm within the vicinity of the specimen surface are found detrimental for fatigue damage. Randomly distributed porosity in SLM parts increases the fatigue scatter significantly [51,55,71].

Microstructure

For the parts having micro-porosity, the porosity-induced fatigue damage remains the dominant parameter. If the porosity is controlled, the next parameter comes into play is microstructure; however, residual stresses, if present, may also overrule the effect of microstructure [44,72]. Microstructure of a material is representative of the manufacturing process and the post-processing treatments which it has gone through. Fine microstructure has more grain boundaries and more resistant to micro-crack initiation due to grain boundaries causing inhibition of crack propagation. The effect of grain size on crack growth has been understood for Al-Mg alloys and wrought alloys [72,73], but its effect in cast and melted alloys is not clear. For eutectic alloys like AlSi12, the microstructural parameters of importance are Al-matrix, amount of AlSi eutectic and the size of dendrites. As the grain size in SLM parts is very fine, it decreases the fracture toughness which may increase the rate of crack propagation.

Residual stresses

Residual stresses are the stress distributions which are already present in a part without application of external forces. Residual stresses are worth considering for fatigue as they affect the same way as that of external loads. They can be beneficial or detrimental for fatigue life depending on their local conditions. A consequence of residual stresses is to increase or decrease the effective stress amplitudes. If these stresses are tensile, they will increase the effective amplitude of applied stresses, which will correspondingly be detrimental to fatigue life. If residual compressive stresses are present, they will decrease the effective amplitude of applied stresses and will be beneficial for fatigue life. Tensile and compressive residual stresses appear in a part simultaneously. In the absence of external loads, they balance each other. So it is always critical to understand their localized effect. If there are compressive residual stresses at locations favorable for fatigue failure balanced by tensile stresses at locations not favorable for fatigue failure, it will be beneficial for fatigue life. Notch roots or other parts having uneven stress distributions are always sensitive spots where crack nucleation is more likely. Residual stresses may be developed by the manufacturing process, heat treatment or in assembly. These stresses can also be induced according to requirement to produce favorable fatigue conditions e.g. by shot-peening. It may also be required to remove residual stresses if unfavorable tensile stresses are present. Some sort of heat treatment may be done to get relief from stresses [74–76]. High cooling rates in SLM process result in large residual stresses which have been highlighted in section 2.1.3.

Surface effects

The factors which can effect positively or negatively the crack nucleation and crack growth related to surface finish are important. These include roughness, surface damage, and surface treatments like anodizing, nitriding, carburizing, decarburizing, cladding or shot-peening. Surface roughness is of significant importance for fatigue crack initiation. Fine surface has a beneficial effect on crack initiation at all stress amplitudes. The maximum difference between the surface profile of the specimen acts as notches, and, therefore, dominant crack initiation sites. Other surface damages are also favorable to crack nucleation. The overall effect of all the factors is to reduce the fatigue life of a part. Surface roughness contours at the specimen act as micro-notches and a favorable source of early crack nucleation which decreases the fatigue life considerably. The effect of surface roughness becomes more prominent in the high cycle and very high cycle fatigue range. The surface roughness obtained from the SLM process is usually in the range of average roughness $R_a = 5\text{-}15$ μm. Such high roughness may decrease the fatigue strength significantly [57].

2.2.3   Crack initiation and crack propagation mechanisms

Under repeated application of load, a fatigue mechanism is initiated by a minor-crack, which grows afterwards till the complete failure of the structure. The crack at the start is usually at microscopic level which grows to macroscopic level and finally causes failure. The life of a component is composed of three stages – crack initiation, crack propagation, and the final failure. The crack initiation phase is usually the largest portion of the total life followed by crack propagation. Some parameters have different nature of effect in the two phases. Fig. 2.6 shows the phases of fatigue life and the corresponding factors of importance as K_t (stress concentration factor), K (stress intensity factor) and K_{IC} (fracture toughness) [74].

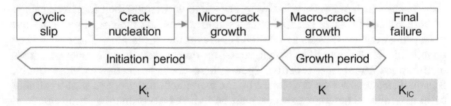

Fig. 2.6:   Different phases of fatigue life of materials [74]

In fatigue testing, specimens are tested under cyclic loading either until a specific number of cycles or until failure. The applied loading can be constant amplitude or variable amplitude and the tests may be performed by controlling stress, total strain or plastic strain. A Woehler curve is determined by performing constant amplitude fatigue tests at several stress states until the specimen fails. Fig. 2.7 shows the load time history of a stress-controlled constant amplitude test using sinusoidal function. Such a fatigue test is characterized by the stress amplitude, mean stress, load ratio and the test frequency. The important parameters for fatigue loading are stress range ($\Delta\sigma$), stress amplitude (σ_a), mean stress (σ_m) and laod ratio (R), and are defined as:

$$\Delta\sigma = \sigma_{max} - \sigma_{min} \tag{2.3}$$

$$\sigma_a = \frac{\Delta\sigma}{2} \tag{2.4}$$

$$\sigma_m = \frac{\sigma_{max} + \sigma_{min}}{2} \tag{2.5}$$

$$R = \frac{\sigma_{min}}{\sigma_{max}} \tag{2.6}$$

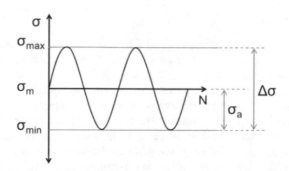

Fig. 2.7:   Characterization of a constant amplitude fatigue test

After performing several fatigue tests at specified stress levels, the resulting data is plotted in the form of a Woehler curve where there exists a log-linear formulation between stress state and fatigue life. This relationship is formulated, according to Basquin [77], as given in eq. 2.7.

$$\sigma_a = \sigma_f' \cdot (2N_f)^b \tag{2.7}$$

with σ_f' being the fatigue coefficient, b as the fatigue exponent and $2N_f$ as the number of load reversals until failure. Woehler curve is characterized by an accompanying scatter, therefore many tests are required to be carried out according to the desired confidence level and probability of failure [78,79]. To find the fatigue strength for a specific fatigue life, staircase method is widely accepted [79].

Current advanced research in fatigue failure focuses on developing improved and time-efficient methodology for fatigue behavior of materials and structures. Though Woehler curve is widely accepted and reliable, it gives no information about the behavior of the material during testing and gives only life until failure. However, if material reaction to external loading is observed, it can give significant information about the mechanisms and microstructural changes in the core of the material. A so-called load increase test has been employed where cyclic load is applied for a fixed number of cycles and increased continuously or step-wise until failure. Such test can give an indication of the range of stress amplitudes which is critical for that material or structure, so that the proceeding constant amplitude tests need to be carried out for that stress range [80]. Additionally, different sensor techniques can be applied to trace the physical and microstructural changes which occur during cyclic loading [81–83]. The details of the procedure are explained in section 0 .

Crack initiation

Crack initiation is an effect produced by slip bands at microscopic level. These slip bands are actually micro-plastic deformations, also called dislocations, acting at the grain or sub-grain level. This may be termed as micro-plasticity as it is limited to a few number of grains. The slip bands are initiated usually at the free-surface, as it offers the minimum resistance as compared to sub-surface grains. Even at the surface, the entire surface is not equally favorable for all the slip systems. The spot having some discontinuity will be subjected to stress concentration, and will be susceptible to fatigue crack initiation. Some other reasons include inhomogeneous stress distribution, surface roughness, corrosion pitting and fretting fatigue [74].

Slip bands are a result of shear stresses. Cyclic shear stresses are not uniformly distributed, they may differ in different grains depending on the size and shape, the orientation in the crystallographic planes and also on the material anisotropy. When a slip appears, a new layer of the material is in free contact to the environment. This new layer is readily covered by oxide layer. The slip occurred during loading also introduces some strain hardening in the slip system. Upon unloading, a shear stress is present in the slip band in a direction opposite to the previous one. However, the process is not fully reversible which causes slip bands. The irreversibility is due to oxide layer and the strain hardening effect. A micro-crack can be initiated on application of a single load cycle which, in subsequent cycles, causes crack extension. Initially it is reversible but, after crossing a few grains, the damage becomes irreversible and the slip bands are then called persistent slip bands (PSBs). These are called persistent because they would appear again if cyclic loading is applied after any surface treatment e.g. electro-polishing [84]. A micro-crack can be an intrusion or an extrusion (Fig. 2.8), depending on the reversed slip; if it occurs at the lower side, it will result in an extrusion. This mechanism also disrupts atomic bonds causing de-cohesion which can be aggravated by aggressive environment [74,85].

Based on this background, fatigue crack initiation can be termed as a surface phenomenon. However, it may occur in the core of the test specimen in several cases e.g. presence of inclusions, micro-pores or internal defects in the core of the specimen. Internal crack initiation may decrease the fatigue life of a component drastically depending on the size of the material inhomogeneity [86,87].

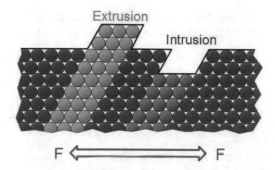

Fig. 2.8:   Extrusion and intrusion mechanisms in fatigue crack initiation [88]

Crack propagation

When the micro-crack penetrates a few grains into the sub-surface, it does not remain dependent on the surface features like roughness or environment. It is then usually a function of material bulk property. Due to the micro-crack, the stress distribution at micro-level is disturbed and the crack tip will experience stress concentration, due to which more slip systems are activated. The direction of micro-crack growth is usually found perpendicular to the loading direction [74].

Rate of crack propagation is an important parameter for calculating the remaining life time of a component in the presence of an existing crack. For fracture mechanics problems, stress intensity factor K, and for cyclic loadings ΔK is the relevant parameter which takes component geometry, crack length and the applied loading into account. The range of stress intensity factor is given by eq. 2.8.

$$\Delta K = \Delta \sigma \cdot Y \cdot \sqrt{\pi \cdot a} \qquad\qquad\qquad\qquad (2.8)$$

where;

   ΔK: Range of stress intensity factor
   $\Delta \sigma$:   Range of applied stress
   Y: Geometry factor
   a: Crack length

Therefore, rate of crack propagation is portrayed in terms of stress intensity factor range. The graph in Fig. 2.9 can be classified into three regions. At low values of stress intensity factor range, the graph takes the form of a vertical asymptote which implies that the applied stresses are small enough to propagate the crack further. So the crack does not propagate in this region corresponding to the threshold stress intensity factor ΔK_{th}. In the middle region, called Paris region, crack propagates at a constant rate in a linear form when the data is presented in double-log form. The rate of crack propagation is given in eq. 2.9 as:

$$\frac{da}{dN} = C \cdot (\Delta K)^m \qquad\qquad (2.9)$$

where;

da/dN:	Rate of crack growth
C:	Material constant
ΔK:	Range of stress intensity factor
m:	Slope of Paris line

As the range of stress intensity factor increases, either by increase in crack length or in applied stress range, the rate of crack propagation does not remain stable, resulting in rapid increase in the crack growth followed by rupture of the specimen. The corresponding stress intensity factor range is termed as critical stress intensity factor range denoted by ΔK_C. For better fatigue performance of a component, a higher threshold stress intensity factor and critical stress intensity factor as well as a relatively small crack growth rate at a given ΔK are required [89,90]

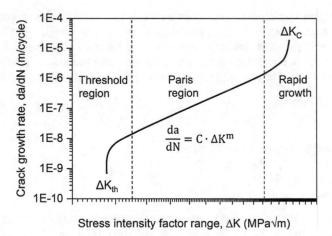

Fig. 2.9:   Representation of fatigue crack growth rate as a function of stress intensity factor range highlighting three distinct regions i.e. threshold region, Paris region and region of rapid growth, adapted from [89]

2.2.4   Fatigue prediction methodologies

The demand for rapid evaluation and prediction methodologies of fatigue performance have risen from the ever-increasing rapid nature of modern manufacturing processes as well as improved material performance. Experimental testing platforms e.g. determination of Woehler curves are usually time- and effort-extensive until the results of fatigue strength are available and could be used for design purposes. Additionally, the produced results are normally representative of ideal loading and environmental conditions. They cannot cope with the complicated load spectra or environmental conditions of onsite applications, which limits the development chain of new materials and structural designs. Integration of modern numerical and stochastic prediction algorithms in support of experimental effort provides a comprehensive platform for rapid characterization and complements the lack of flexibility of purely experimental setups. Modeling of fatigue strength has two predominant routes which are either establishment of relation between crack length and cycles to failure [91], or the relation between other damage parameters like plastic strain amplitude, change in temperature or change in electrical resistance with respect to number of cycles during the course of fatigue life [80,82]. The existing approaches used for prediction of fatigue performance are discussed below.

A   Fracture mechanics-based approaches

This approach is based on the concept of small scale yielding that is concerned with the formation of a local plastic yielding zone ahead of a crack tip even when the material is not yielding on a macro-scale. The phenomenon is widely valid for materials with low and intermediate strength. In this regard, the mechanics of crack growth follows two broad classifications based on scale of fatigue cracks i.e. long cracks and short cracks.

Fatigue crack growth of long cracks

Long cracks are classified as cracks that are long enough to develop crack closure. Such cracks can be governed by Kitagawa-Takahashi model of eq. 2.10. The force driving crack propagation, on this scale will be the accumulation of plastic damage ahead of the crack tip reaching a critical level. Typical behavior of crack propagation rate can be described with respect to the applied stress intensity factor range ΔK through Paris expression in eq. 2.11 [74,92,93].

$$\sigma_{th} = \frac{K_{th}}{Y\sqrt{\pi \cdot a}} \tag{2.10}$$

where σ_{th} is the threshold stress, K_{th} is the threshold stress intensity factor, Y is a geometry parameter, and a is the crack length.

$$\frac{da}{dN} = C \cdot (\Delta K)^m \tag{2.11}$$

where C and m are material parameters and ΔK is the stress intensity factor range.

Since long fatigue crack growth is likely to develop closures and induce crack blunting, the nominal stress intensity factor range does not provide accurate effect over the crack propagation rate. For this reason, correction needs to be applied with respect to mean stress and fatigue strength coefficient which yields an effective stress intensity factor for the initial propagation in eq. 2.12 and for subsequent cycles in eq. 2.13 accounting for variation subject to fatigue strength and applied mean stress [91].

$$(\Delta K_{eff})_0 = \Delta K[1 - \frac{1}{2}(\frac{\Delta K_{th}}{\Delta K})^2] \tag{2.12}$$

$$\Delta K_{eff} = \frac{(\Delta K_{eff})_0}{[1 - (\sigma_m/\sigma'_f)]} \tag{2.13}$$

where ΔK_{eff} is the effective stress intensity factor range, σ_m is the mean stress, and σ'_f is the fatigue strength coefficient.

Fatigue crack growth of short cracks

On a microstructural scale, a single slip band corresponding to the representative length of the microstructure is equivalent to the minimum required cyclic plastic zone for the crack to grow continuously. In this account, it represents the crack potential to overcome microstructural barriers. Thus the problem is confined to a sub-layer of the microstructure where micro-stresses have overcome the yield strength of the bulk materials. Hence, crack progression per cycle comes to be dependent on the characteristic layer thickness as well as micro-yielding properties, as shown in eq. 2.14 [91].

$$\frac{da}{dN} = 2\delta^* [\frac{\Delta\sigma_{ym}(\Delta\varepsilon_b - \frac{\Delta\sigma_{ym}}{E}) + (\Delta K^2/\pi E)}{4 \cdot \sigma'_f \cdot \varepsilon'_f}]^{1/(b'+c')} \tag{2.14}$$

where δ^* is the characteristic layer thickness, σ_{ym} the microscopic yield stress, ε_b strain of the bulk material, ε'_f the ductility coefficient, E the elasticity modulus, and b' and c' are material strength exponents.

B   Damage-based approaches

The accumulation of fatigue damage follows a continuum plasticity scheme where the stress-strain relation is depicted on number of load reversals. This brings into consideration phenomena like Bauschinger effect as well as cyclic ratcheting and hysteresis relaxation. When bulk material flows plastically under loading, distribution of micro-stresses around dislocations and dislocation cells induces unstable cyclic response highlighted in the plateau of the plastic strain amplitude, temperature and electric resistance. Non-linearity of response is augmented more when specimen under loading contains porosity or non-metallic inclusions [80]. Damage-based fatigue characterization techniques can rely on the applied stresses or the induced strains which are described below.

Stress-based approach

The number of cycles to failure under fatigue loading can be related to damage parameters or directly to the stress which has induced fatigue damage within a specimen. In this scheme, the number of cycles is related to the applied stress amplitude σ_a by the corresponding fatigue strength coefficient σ_f' and fatigue strength exponent b according to eq. 2.15 to 2.17. This scheme is specifically suited for stress ranges resulting in low values of plastic strains typically in the range of HCF and VHCF where Basquin relation can be applied, as the applied loads are dominantly in elastic range. Such rapid representation is of high interest in industrial applications since it treats fatigue damage from a macro-scale point of view [91].

$$\sigma_a = \frac{\Delta\sigma}{2} = \sigma_f' \, (2N_f)^b \tag{2.15}$$

$$b = \frac{-n'}{1 + 5n'} \tag{2.16}$$

$$\frac{\Delta\sigma}{2} = (\sigma_f' - \sigma_m)(2N_f)^b \tag{2.17}$$

where σ_a is the applied stress amplitude, $\Delta\sigma$ is the stress range, σ_f' is the fatigue strength coefficient, N_f is number of cycles to failure, b is the fatigue strength exponent, n' is the cyclic hardening exponent, and σ_m is the mean stress.

Strain-based approach

For low cycle fatigue applications where plastic strain amplitudes are significant, distribution of micro-stresses leads to saturation and early crack initiation where compliance leads to increased strain response. Thus control of strain during the test provides more stable cyclic loading condition for specimen under investigation. In contrast to stress-based approach, no connection can be found between quasistatic fracture strain and cyclic fracture strain; however, a strain-based analysis provides an accurate representation for metals cyclically loaded beyond elastic limit. Thus low cycle fatigue can be better represented by Coffin-Manson relation as described in eq. 2.18, where the fatigue ductility exponent, given in eq. 2.19, is of utmost importance. Combining the LCF and HCF ranges, the accumulative equation can be found as described in eq. 2.20 [91].

$$\frac{\Delta \varepsilon_p}{2} = \varepsilon_f' \, (2N_f)^c \tag{2.18}$$

$$c = \frac{-1}{1 + 5n'} \tag{2.19}$$

$$\frac{\Delta \varepsilon}{2} = \frac{\Delta \varepsilon_e}{2} + \frac{\Delta \varepsilon_p}{2} = \frac{\sigma_f'}{E} (2N_f)^b + \varepsilon_f' \, (2N_f)^c \tag{2.20}$$

where $\Delta \varepsilon_p$ is the plastic strain range, ε_f' the fatigue ductility coefficient, and c the fatigue ductility exponent.

Role of statistical techniques

Accumulation of knowledge in fatigue research has proven the fatigue damage as a multi-factor process, dependent on dislocation mechanics up in scale to macro-defects as well as surface and loading conditions. This results in a scatter of fatigue strength values whether experimental or computational, since it is not possible to reproduce the same conditions exactly for every component. The variability nature of fatigue strength parameters offers stochastic modeling as a complementary approach in modeling and prediction applications [94]. A well-established stochastic fatigue strength model is the weakest-link theory relying on Weibull's continuous probability density function presented in eq. 2.21. The model relies on Weibull parameters as a defect distribution character and a material property. The load is varied either statistically or numerically until a certain failure probability is reached. The targeted failure probability depends on the base of Weibull's exponent. In a wider view, the modeling procedure includes an application of a failure criterion e.g. Dang-Van HCF criterion found in eq. 2.22. The model proved adequate for prediction of fatigue strength on various scales and materials [91,95,96].

$$f(x; \alpha, \beta) = \begin{cases} \alpha \cdot \beta \cdot x^{\beta-1} \cdot e^{-\alpha x^\beta}, & x > 0 \\ 0, & \text{else} \end{cases} \tag{2.21}$$

$$\sigma_{DV} = \tau_{max} + \varphi \cdot P_{max} \tag{2.22}$$

where σ_{DV} is Dang-Van equivalent stress at any time point, τ_{max} is the maximum shear stress, φ is the sensitivity factor, α and β are the Weibull's parameters, and P_{max} is the maximum hydrostatic pressure

Probabilistic fatigue crack growth

Probability and statistics have been extensively applied to model fatigue crack growth on various scales since the micro-scale associated interactions are cumbersome to realize deterministically. For realization of the problem, damage zone ahead of crack tip is limited to a reversed plastic zone beyond which damage is not significant. Element meshing ahead of crack tip should anticipate not to exceed fatigue strength or ductility coefficient. Utilizing a small scale yielding concept, the stress intensity factor range ΔK can be related to critical values of stress and strain according to eq. 2.23. The value of x^* is usually in the order of the grain size of polycrystalline materials. The random crack propagation rate will then be realized in eq. 2.24 [91].

$$x^* = \frac{\Delta K_c^{\ 2}}{I_n' \cdot \varepsilon_f' \cdot \sigma_f' \cdot E} \tag{2.23}$$

$$\left\langle \frac{da}{dN} \right\rangle = \frac{i x^*}{n} \tag{2.24}$$

where I_n' is a non-dimensional parameter of fatigue strength exponent n', and n is the number of cycles.

Role of finite element modeling

The time dependency of the stress and fatigue strength under cyclic loading can be reduced to partial equations describing the stress displacement relations within a specimen or a component. Such mathematical formulation cannot be realized in an analytical form. On the contrary, finite element modeling (FEM) provides a numerical approximation of this load-displacement time dependency [97]. With such a capability, stress and plastic strain distribution within a component can be realized to establish expressions about the effect of geometrical non-linearities (such as notches, porosity, and surface roughness) on the expected fatigue strength [98]. Additionally, the problem of cyclic response instability can be addressed in FEM via coupling of Fourier series and Newton-Raphson scheme. When FEM results are coupled with a proper fatigue failure criterion in post-processing, fatigue life time prediction could be made with considerable accuracy. Prediction accuracy can be enhanced based on non-destructive testing (NDT) techniques such as computed tomography (CT).

2.3   Process-property relationships in SLM parts

2.3.1   Physical properties

Structural part properties – relative density, microstructure and residual stresses – are important metrics to be considered when ensuring quality and repeatability of the SLM process. Relative density has a significant effect on many material properties like hardness, quasistatic behavior, fatigue behavior as well as the corresponding failure mechanisms. The densification mechanism is controlled by the SLM processing parameters especially beam intensity, scanning speed and hatch distance which effect the rate of cooling. Currently most studies report a relative density of more than 99%. The most important parameter for achieving a dense material is an appropriate laser energy density imparted to the powder material. In general, low laser power, high scan speed, large hatch distance and large layer thickness create bonding defects in the material. Therefore, process optimization for each material or alloy should be carried out for the selection of an optimal energy density [99–102].

There are certain phenomena in SLM process responsible for remnant porosity. One of these is the so-called balling effect which occurs due to the tendency of melt track to shrink. It results due to surface tension to decrease the surface energy. The unbalance of forces from inside and outside of the melt pool induces torque. Depending on the intensity of the torque, the heavier melt will concentrate at the periphery of the melt, and gas or vacuum will be formed at the center. The balling phenomenon is detrimental to the quality of SLM parts in terms of pore formation between these balls. These metallic balls may also distract the uniformity of powder layer spread by roller or blade which can further increase the issue of porosity as well as contribute in rough surface of the SLM parts. Many factors are important to be considered to attenuate the balling effects. Atmospheric oxygen has an important role in the severity of the balling, as the oxygen content in the SLM part increases as the atmospheric oxygen content increases. Increasing the scan speed decreases the wettability and causes discontinuous melt tracks resulting in balling. The same effect is produced by low laser power. Much higher scan speeds will be deleterious not only for relative density but also for surface finish. Large layer thickness also favors balling effect because the contact area between the melt pool and the previously deposited substrate decreases, thereby decreasing the wetting area, it becomes insufficient for supporting a big melt pool resulting in dissociation of melt pool into balls. Therefore, a careful selection of the process parameters is an important factor for quality parts, and the size of the used powder should be conforming to the layer thickness [103,104]. The shape of the pores can be complex and can have concave or convex fronts [105]. Gas porosity is another phenomenon responsible for inducing pores in SLM parts. Some of this porosity can also be inherent powder porosity as well as the oxides existing in the powder material if it still contains moisture which may not be completely escaped during cooling. Though inert gas is used inside the chamber; some gas can be entrapped in the material during solidification [106]. Therefore, care should be taken in using the quality of the input powder and some mechanisms for de-gassing should be developed.

The unique microstructure obtained by the SLM process is representative of the specific melting conditions and can be controlled by varying the process parameters. Short exposure times and high thermal gradients lead to fine microstructure and development of residual stresses. Typical microstructure obtained by the SLM process is columnar / lamellar in nature where the needles grow within a powder bead [42,107]. Core of the melt pool contains a finer microstructure, but it changes to a coarser microstructure at the melt pool boundary because of the more exposure time due to double tracking region by two adjacent hatch tracks.

The microstructure obtained by the SLM process is usually of metastable nature and may change depending on the post-process heat treatment [47]. Fig. 2.10 portrays the effect of post-process heat treatment on the evolution of microstructure in SLM-manufactured AlSi12 alloy. The alloy was selective laser melted and then subjected to isothermal annealing treatment at 473 K (Fig. 2.10a), 573 K (Fig. 2.10b), 673 K (Fig. 2.10c), and 723 K (Fig. 2.10d). The microstructural features show that, after annealing treatment, cellular dendrites of the microstructure change to a mixed microstructure as the annealing temperature increases. The temperature does have an influence on the Si particles especially in the melt boundaries where these act as a starting site for the formation of Si particles. The insoluble Si is ejected from the Al matrix and is spread around these sites, as shown in the diagram below as a function of temperature. Size of the Si particles also is influenced by the temperature of the annealing. These Si particles act as a source of reinforcement [47,108].

Microstructure of the SLM parts exhibits anisotropy [42,109]; the microstructure perpendicular to the build direction usually represents the used scan strategy, and the width of the separate melt tracks corresponds to the hatch spacing. Alternating direction of scanning results in herringbone pattern, which infers that the direction of heat transfer will determine the orientation of the major microstructural features

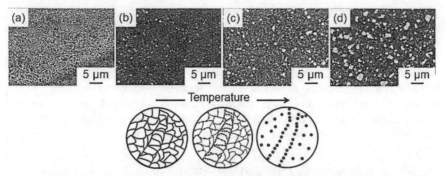

Fig. 2.10:   Microstructural evolution of AlSi12 alloy after annealing heat treatment at:
473 K (a), 573 K (b), 673 K (c), and 723 K (d), followed by a schematic
representation of Si ejection, adapted from [47]

In the building direction, grains appear in elongated / dendritic pattern epitaxially
deposited in the substrate structure which indicates that the melt pool obtained in
the SLM process is significantly higher than the layer thickness [42]. Therefore, it
suggests that uniform scanning strategy results in anisotropic microstructure and
alternating, bi-directional, and continuously rotating scanning strategy be used if
relatively uniform microstructures are required.

One of the potential applications of SLM process is to manufacture thin-walled
and functionally-graded structures, therefore, the localized effects should be taken
as a competitive advantage of the process for such structures. To exploit this
feature, the role of individual process parameters in controlling the microstructure
needs to be understood. Studies have shown that decreasing the scanning speed
results in coarser and elongated microstructures which are in alignment with the
building direction. Laser power also plays an important role in defining the
microstructure. Higher laser power enhances the melt pool, thereby increasing the
wettability resulting in relatively coarse grains. Multiple laser sources can be
installed in an SLM machine to produce different microstructures at different
locations of the parts resulting in functionally-graded structures [110,111].

Another concern in SLM parts is residual stresses. Several factors contribute to
their development during the process, the most important being the process nature
itself introducing temperature gradient mechanism [59]. Thermal expansions and
contractions experienced by the substrate with high rate generate thermal stresses
depending on the cooling rate. High tensile stresses have been measured at the
upper region of the built part. Scanning strategy has also been found influencing
the residual stresses with island scanning at small island sizes giving the optimal
results due to reduced anisotropy of the melt pool as compared to longitudinal line
scanning [42]. Base plate has an important influence on the residual stresses in

terms of its thickness as well as temperature. Thick base plate will offer smaller deformation during part removal resulting in smaller residual stresses [59]. Heating the base plate during the manufacturing process decreases the thermal gradient which relaxes the induced stresses [51]. In that case, the height of the substrate above the base plate may also have an influence on the residual stresses. With increasing substrate height, the thermal gradient would increase resulting in increased residual stresses. Additionally, material properties determine the magnitude of residual stresses. Materials with high yield strength favor higher residual stresses in the substrate due to the dependence of gradient development on the plastic deformation behavior during solidification. Molten material in the current layer experiences restrictions to expansion due to previously existing layer, which gives rise to development of elastic compressive strains. The current layer can be compressed plastically when its yield point is reached. The plastically-compressed layer will start shrinking resulting in higher residual stresses [59]. In the absence of the objective of manufacturing localized properties, the determination of bulk material properties should be carried out after a post-build stress-relief heat treatment with material-dependent parameters.

Another important parameter in determining the quality of SLM parts is surface finish. In SLM process, surface quality is influenced by a number of parameters. One of the major phenomena is the processing technique itself where the part is manufactured layer-wise. During conversion of the CAD model to two-dimensional slices, a so-called stair step effect is observed due to the approximation from 3D to 2D horizontal layers [112]. Larger layer and beam thicknesses employed to increase the productivity of the process affect the surface quality [113]. Distribution of the powder particle size is important in determining the resulting surface quality, with finer and smaller particles resulting in better surface finish [114]. Another process-specific phenomenon is balling effect which occurs due to the surface tension and the molten material does not stick to the previously deposited substrate resulting in bead-like rough surface [104,115]. Surface re-melting after each layer can not only improve the surface quality but also higher densities can be achieved [42,56,116].

2.3.2 Quasistatic properties

Quasistatic tensile strength of the SLM parts is a function of the part properties obtained by different processing parameters. Surface roughness of the parts does not significantly affect the bulk material properties like tensile strength, yield strength and toughness. Almost fully dense parts with optimal properties have been found to have quasistatic strength values reaching or even surpassing those of wrought material for the investigated materials – steel [117,118], titanium [55,119,120], aluminum [47,51,52]. The relatively high strength is attributed to the very high thermal gradients in the process resulting in a fine microstructure. Such fine microstructure is also responsible for decrease in ductility which is sufficiently less as compared to that in wrought materials. Coarse microstructures obtained by heating the base plate decrease the tensile strength and increase the fracture strain according to Hall-Petch effect [51].

Variations in mechanical properties as a function of build orientation have been observed. Buchbinder et al. [121] have investigated the mechanical properties of AlSi10Mg alloy at 0°, 45° and 90° to the loading direction and found that loading direction parallel to the build orientation gives the highest tensile strength and the perpendicular direction results in the least strength. Manfredi et al. [122] have investigated the specimens built parallel and perpendicular to the base plate for the same alloy. They also investigated the orthotropy at three different orientations in the plane of the base plate. No difference in the mechanical properties was found for the specimens built at different orientations of the same plane. However, for the specimens built perpendicular to the base plate (parallel to the loading direction), fracture strain was sufficiently reduced with no difference in tensile strength.

Suryawanshi et al. [123] have recently investigated the quasistatic tensile properties of AlSi12 alloy manufactured with SLM in loading direction parallel and perpendicular to build direction, and the results of tensile strength, yield strength and fracture strain are plotted in Fig. 2.11, where AS represents as-built specimens from SLM process, HS the specimens after heat treatment at 573 K and 6 hrs, ‖ shows the loading direction in parallel to build direction and ⊥ in the transverse direction. Much higher tensile and yield strengths, as compared to cast alloy, have been obtained with a corresponding decrease in fracture strain. Heat treatment at 573 K has resulted in decreased tensile strength and yield strength values with a corresponding increase in fracture strain, which is valid both for parallel and perpendicular directions. The influence of build direction is not found significant for tensile strength and yield strength, but its effect is consistent for fracture strain where parallel build direction has resulted in higher fracture strain as compared with those built at perpendicular direction.

Hanzl et al. [118] have investigated the influence for Ti alloy and results conforming to the previous ones were obtained for the tensile strength. However, in this study, fracture strain was also higher for the vertically built direction (loading direction parallel to building direction). The specimens built at oblique angles were found to possess best combination of strength and toughness. Study of Shunmugavel et al. [120] on Ti alloys confirms the higher tensile strength in the building direction parallel to the loading direction; however, no difference in the fracture strain for the two orientations was found in this study. A study by Delgado et al. [124] on iron-based materials have found no significant influence of build direction on the tensile properties. The studies available to date do not agree about the influence of orientation on the strength, and needs to be understood together with the influence of other factors on mechanical properties.

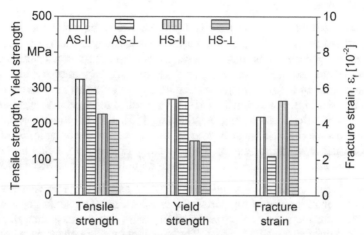

Fig. 2.11: Influence of build direction and heat treatment of quasistatic tensile properties of AlSi12 alloy, AS represents as-built specimens, HS heat treated specimens, ‖ and ⊥ represent parallel and perpendicular build direction respectively, adapted from [123]

Another important parameter of consideration is the influence of build rate on the mechanical strength. Though it is required that the parts are having full density for better mechanical properties, one of the potential advantages of SLM process is its inherent capability to manufacture porous structures which can favor design for light-weight manufacturing. Current studies mostly focus on process optimization for full density; however, it is required that the possibility of manufacturing selective porous structures be investigated in terms of their structural strength to favor resource efficiency and economy.

2.3.3   Fatigue properties

Many of the potential applications of SLM parts are in aerospace, automotive and medical implant industries which require, additional to quasistatic strength, reliable fatigue data for component design. For qualification of the SLM process for industrial and medical applications, fatigue performance of the parts should be investigated and quantified. Fatigue behavior of the SLM parts, and of additively-manufactured parts in general, is a topic which has not yet been addressed in detail. Only some studies are available recently which have investigated the fatigue-related mechanisms of SLM parts. Several of the SLM process and part parameters are important in determining the fatigue behavior of SLM parts. The important part parameters are surface roughness, material defects, microstructure and residual stresses.

Rough surfaces obtained from the SLM process are detrimental for the fatigue strength of the parts [55,125]. Rough surfaces in the range of tens of microns of maximum roughness act as micro-notches causing an early crack initiation and impaired fatigue strength. However, post-processing can improve the surface finish and result in fatigue strength almost comparable to that of conventionally-manufactured alloys. Studies on Ti-6Al-4V showed an increase in fatigue strength of about 100% after shot-peening or polishing [55]. It is not always practical to improve the surface finish by post-process machining for the complex structures, therefore surface modifications by sand-blasting or other appropriate chemical machining needs to be explored for process and application-specific conditions.

Process-induced imperfections, bonding defects and gas porosity, are critical parameters for reliability of the parts in cyclic loading. The small pores are not very critical for the quasistatic strength of the parts, but act as source of fatigue crack initiation under cyclic loading. The distribution of remnant pores in SLM parts is not uniform which results in non-uniform fatigue life at a specific stress i.e. they have high fatigue scatter. It has been investigated for some of the materials like Ti-6Al-4V [125,126,128], AlSi alloy [54] and steels [43,57,127]. The general observation is that there exist different crack initiation modes in SLM parts which are surface crack initiation and crack initiation from bonding defects, as summarized in Fig. 2.12 for Ti-6Al-4V. For the as-built condition, having surface roughness, the fatigue strength is small, but reliable; however, after surface treatment and shot-peening treatment, fatigue strength increased, but also decreasing the corresponding reliability which is attributed to crack initiation mechanisms.

If there are no significant defects in the vicinity of the contour, the fatigue crack initiates from the boundary due to surface weakness. Surface cracks from smooth surfaces usually result in fatigue lives corresponding to the material strength. If there is a defect of significant size, and that too near to boundary, there is a high probability of crack initiation from that defect. Then the size and location of the crack-initiating defect would determine the final life of the parts [129]. Therefore, to ensure reliability of SLM parts in fatigue applications, it is important to control the remnant porosity in terms of its size and location.

If the surface and internal defects are controlled, next phenomena come into play are microstructure and residual stresses. Fatigue strength of the SLM parts is sufficiently reduced, as compared to conventionally-manufactured parts, which is

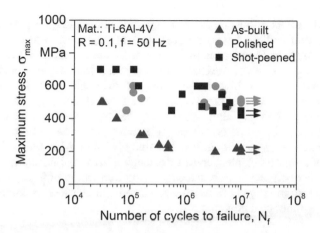

Fig. 2.12: Woehler curve for Ti-6Al-4V showing influence of polishing and shot-peening, adapted from [55]

due to surface effects; however, if the effect of microstructure is to be investigated, surface effects need to be controlled. Therefore, a review is made for the studies which have investigated polished specimens so that microstructural effects can be investigated and compared with wrought alloys. The characteristic microstructure obtained in the SLM process is discussed in section 2.3.1. Though material-dependent, it generally consists of ultra-fine columnar grains which are oriented in the direction of build. This microstructure results in fatigue strength comparable to that of wrought alloys [127,130], and higher if other features like remnant porosity is controlled. The fine microstructure serves as hindrance to dislocation motion improving the fatigue strength. It has been demonstrated by Rafi et al. [128] for polished Ti-6Al-4V specimens, and Wycisk et al. [131] for polished as well as post-process hot isostatic pressed (HIP) for Ti-6Al-4V specimens. HIP process removes the remnant pores and is, therefore, the best case for investigating the effect of microstructure, developed in SLM process, on fatigue behavior. The fatigue strength in that case was observed to be significantly high.

Post-process heat treatments have been investigated by some other researchers. Leuders et al. [132] have carried out heat treatments on Ti-6Al-4V at 800 °C, 1050 °C and HIP treatment at 920 °C which increased the fatigue life four to ten times as compared to as-built conditions. These improvements are representative of the microstructural changes in the heat treatment process. The corresponding residual stresses were also reduced. The as-built specimens had tensile residual stresses of 90-265 MPa in x-direction and 235-775 MPa in y-direction which was reduced to maximum of 10 MPa after 800 °C heat treatment. Similar observations regarding the effect of heat treatment are mentioned in [133,134]. Selective laser melted aluminum alloys have been a subject of fatigue even less than titanium alloys. One study by Brandl et al. [44] has investigated the effect of heat treatments on the fatigue properties of AlSi10Mg and found that the fatigue strength of SLM-manufactured alloy is higher than that of reference material. They have investigated the effect of solution heat treatment and artificial aging and found that peak-hardening had a high effect on fatigue coefficients because of homogenization of microstructure and spherodization of Si particles resulting in ductile fracture behavior as well. Another study [135] on AlSi10Mg alloy presents the effect of solution treatment and artificial aging in the as-built surface condition. A coarser microstructure was obtained with agglomeration of Si particles, and the overall increase in fatigue life was observed to be three times as compared to the specimens without heat treatment.

2.3.4   Crack propagation properties

A few studies regarding the investigation of fatigue crack propagation of SLM-manufactured materials were presented recently which address the issue in steel and titanium alloys. Riemer et al. [133] have reported fatigue crack growth behavior of 316L stainless steel in three configurations – as-built, heat treated at 650 °C for 2 hrs, and hot isostatic pressed at 1150 °C and 1000 bar. As-built specimens and those heat treated at 650 °C showed similar crack growth behavior and threshold stress intensity factor of 3 MPa√m, which increased to about 5 MPa√m after HIP treatment. This increase can be attributed to the finer grains in the as-built condition and coarsened grains after HIP treatment. Fatigue crack growth resistance for 316L steel was also studied by Okyar et al. [136] in four-point bending tests. They have found that porosity has drastically affected the fatigue crack growth resistance. Riemer et al. [133] have also investigated Ti-6Al-4V for as-built, heat treated at 800 °C and 1050 °C as well as after HIP treatment. As-built specimens had threshold value of stress intensity factor range of less than 2 MPa√m which is about 4 MPa√m for the conventionally-manufactured alloy. This threshold value could be increased by heat treatment temperature and reached too little above 4 MPa√m for HIP and heat treated at 1050 °C. Ti-6Al-4V was also investigated by Cain et al. [137] for the as-built, stress-relief treatment at 650 °C and annealing heat treatment at 890 °C specimens. As-built condition was found to have the fastest crack propagation rate and higher scatter. The crack growth rates decreased for the other two conditions with reduced scatter. Edwards and Ramulu [138] have investigated the fracture toughness and fatigue crack growth of Ti-6Al-4V in three orientations. Specimens built parallel to platform gave the best fracture toughness value and is still lower than that of cast or wrought alloy which is attributed to the martensitic microstructure of the as-built condition. No significant difference in the values of threshold stress intensity factor range was observed for the three orientations. No studies related to fatigue crack growth investigations was found in open literature for SLM-manufactured aluminum alloys.

2.3.5   Inferences and objectives

Thorough review of the available literature in the field reveals that a number of studies are available regarding SLM processing of different alloys. However, these studies focus on the process development for particular alloys. Processing parameters – laser power, scan speed, hatch spacing, layer thickness – have been investigated with an objective to achieve maximum relative density of the parts. The achieved relative density approaches up to 99.5% with a small remnant porosity. The formation of this porosity and its effect needs to be understood in detail. Fig. 2.13 outlines the objectives of this study consisting of the complete process chain. One of the objectives of this study is to explore how this remnant porosity can be eliminated or reduced by in-process manipulations as well as post-process steps. After achieving the maximum density, the role of these remnant defects in static and cyclic applications should be characterized and taken into account. Therefore, the size and distribution of the remnant porosity will be investigated in terms of its magnitude of effect in service applications. For distribution of the internal defects, three-dimensional radiographic technique will be used so that complete profile of the volumetric defects can be detected as compared to two-dimensional metallographic techniques where only sectional properties are obtained. Such data from non-destructive defect detection will be analyzed by finite element methods for their effect on properties. Further, recommendations will be made to optimize the SLM process in a way so that their effect can be minimized.

Microstructure of the manufactured alloys gives important information about the potential properties. Several studies show that SLM results in fine microstructures due to very high cooling rates in the process. Several researchers have carried out different post-process heat treatments to change the microstructure resulting in modified mechanical properties. However, SLM process presents a possibility to control the microstructure during the process. In this study, it is aimed at to investigate the SLM processing parameters so that the cooling rates can be controlled resulting in modified microstructures. Microstructural parameters have a corresponding effect on mechanical properties. Quasistatic mechanical properties have been reported at par with those of conventionally-manufactured alloys for some alloys. For AlSi12 alloy, SLM process will be optimized for mechanical performance. The effect of processing parameters on the part properties and, in turn, mechanical properties will be investigated. Here the role of process-induced residual stresses will also be taken into account. Additionally, ductility of the SLM parts has been reported to be low due to fine microstructures. It is aimed to modify the microstructure which can increase their ductility.

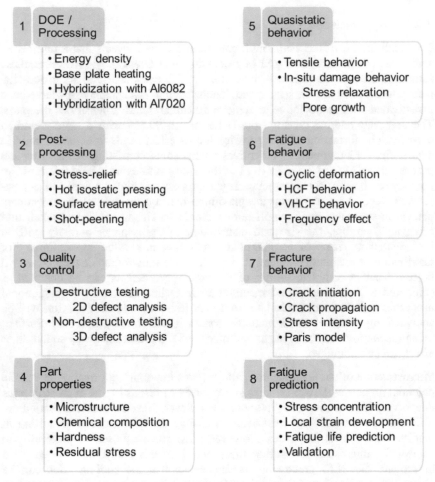

Fig. 2.13: Investigation framework showing the objectives of this study in the process chain

One of the potential benefits of SLM process is to develop porous structures so that objective of light-weight design can be achieved. For several structures, only static strength is of concern, therefore, the mechanical properties of specimens manufactured with reduced laser energy input will be investigated to explore the extent of reduction in mechanical properties as compared to the reduction in laser energy density, which will be helpful in considerations regarding manufacturing of selective porous structures for non-cyclic applications.

As several of the potential applications of the SLM process are in automotive, aerospace and medical industries where cyclic loads are applied, it is important that the complete property profile is available for the designers. Therefore, one of the major objectives of this study is to explore the fatigue behavior of SLM parts. Determination of fatigue behavior has always been an expensive task, as many fatigue tests are required for generating a Woehler curve. Therefore, one of the objectives of this study is to reduce the time for process optimization with an objective criterion of fatigue performance. For this purpose, a combination of so-called load increase tests and constant amplitude tests will be used to reduce the testing efforts. This combination will make it possible to decrease the testing effort by reducing the number of potential parameters for the complete Woehler curve.

Many of the functional parts in automotive and aerospace industries are subjected to load cycles sufficiently above 10^7 cycles. The small existing literature does not make it possible to design the parts for applications in very high cycle fatigue range. The so-called concept of "fatigue limit" is no more valid, as it has been observed that the materials degrade at stresses even below this fatigue limit and the course of the Woehler curve is material dependent [70,139,140]. Therefore, it is required to have reliable fatigue properties in the VHCF range to allow reliable part design in that range. For the applications where fatigue failure is dominated by crack growth mechanism, the crack propagation rates depending on applied stress intensity factors should be investigated so that the remaining fatigue life of a structure can be predicted reliably. Therefore, one of the objectives of this study is to investigate the behavior of fatigue crack propagation and to determine the threshold and critical values of stress intensity factors for different parameters of the SLM process.

SLM process has competitive advantage for the structures which are having intricate shapes which are difficult or cost-extensive to be manufactured by conventional methods. This is particularly relevant for tools and molds which have complex contours. However, complexity of these structures is of partial nature with major parts having geometry which can be manufactured conventionally. Manufacturing of these structures solely by additive manufacturing will hinder the cost-competitiveness. These parts can be manufactured by hybrid techniques i.e. by combining conventional and additive manufacturing. This study will investigate the feasibility of manufacturing these parts by hybrid technique, exploring SLM parameters for optimal joining strength. The corresponding mechanical properties – quasistatic, HCF and VHCF behavior – will be investigated and a comparison will be made with the properties of parts manufactured solely by SLM process.

For utilization of the process in production of small batch novel designs, it is required that the fatigue performance for specific designs and material conditions can be reliably predicted. As far as the process complexity is concerned, fatigue strength is subject to significant variation as a result of minor variation of production route. The ability to link such process-induced changes to final fatigue strength, in a reasonably timely manner, would be beneficial for product design, especially in structural load bearing applications. The interactive variation of fatigue strength between induced structural defects and plastic microstructure-related properties envisage that either a fracture mechanics or plastic damage-based approach can be used for fatigue modeling per respective fatigue attribute. Fracture mechanics models which depend on the propagation of cracks from internal porosity can be used to integrate a component's cycles to failure based on various propagation modes of a crack. This should judge explicitly effects of porosity on the respective fatigue life. At the same time, it considers plasticity on a small scale at the crack tip. Macro-scale plasticity will be utilized to study the effects of microstructure and related plastic properties. Comparison of both groups of modeling results should be a final criterion on determination of the dominant damage model in SLM-manufactured aluminum parts.

2.4   Aluminum-Silicon alloys

2.4.1   Physical properties

Aluminum is the second most abundant metallic material having unique properties of light-weight (one-third that of steel), corrosion resistance, wear resistance, flowability and medium strength. Due to these attractive properties, aluminum and its alloys are used widely in aerospace, automotive as well as construction industries. Pure aluminum is characterized by the physical properties as given in Table 2.1. Aluminum and its alloys can be manufactured for a broad range of strength, from purely ductile (commercially pure aluminum) to sufficiently strong alloys with tensile strength ranging to 700 MPa (Al 7075-T6). They can be used for cryogenic temperatures, as they preserve their strength at low temperatures. High thermal conductivity of aluminum makes it possible to be applied for structures where the transfer of thermal energy is required e.g. in engine components, turbine blades and heat exchangers in several industries. Aluminum is characterized by high reflectivity due to which it can be used as an impediment to thermal radiation e.g. in automotive heat shields. Aluminum has good finishability so that no protective layers are required for many of the applications [141,142].

Si, Mg, Cu and Zn are the alloying elements most commonly used to form Al-alloys. Aluminum-silicon alloys are classified as cast or wrought alloys. Cast alloys are denominated as 4xx and wrought alloys as 4xxx by Aluminum Association. If copper and / or magnesium are added, they are classified as 3xxx alloys. Wrought alloys 4xxx can be manufactured by welding and forging having medium strength, low melting point, narrow freezing range and good flowability.

Table 2.1:   Selected physical properties of pure aluminum [143]

Density	2.70 g/cm^3
Melting point	660 °C
Specific heat	0.91 kJ/kg·K
Thermal conductivity	0.5 cal·s^{-1}cm^{-1}K^{-1}
Reflectivity	90%

Cast aluminum-silicon alloys can have Si content of 5-23% by weight. Use of aluminum alloys as structural materials is influenced by their physical and mechanical properties which depend on their chemical composition. Generally, aluminum alloys are characterized as light-weight as well as having high specific strength i.e. strength as a function of density as compared to other cast alloys. The properties of aluminum alloys highly depend on the microstructural features which are attributed to the chemical composition as well as the processing route. The properties of a particular alloy depend on the physical properties of the individual constituents i.e. relative distribution of the Al matrix and the Si crystals, their volume fraction in the mixture as well as morphological distribution and size [141,143].

Aluminum alloys are important due to their good corrosion resistance. Exposure to the atmosphere produces an invisibly thin (2-3 nm) oxide film which acts as a barrier to further oxidation. The material remains corrosion-protected unless it is exposed to conditions which destroy this protective oxide layer which is stable usually in solutions of pH range 4.5-8.5. Aluminum is also protective against many acids [143].

2.4.2   Phase diagram

Aluminum-silicon forms a eutectic system with a low solubility of Si in Al and Al in Si. Besides melting of pure Al and Si, eutectic transformation from liquid solution to solid solution occurs at 577 ± 1 °C and 12.2 at.-% or 12.6 wt.-% silicon.

The eutectic reaction may be suppressed by about 10 °C if high cooling rates are involved. An exemplary phase diagram is shown in Fig. 2.14 and can be represented by eq. 2.25.

$$L = \alpha + \beta \qquad\qquad\qquad\qquad\qquad\qquad\qquad\qquad (2.25)$$

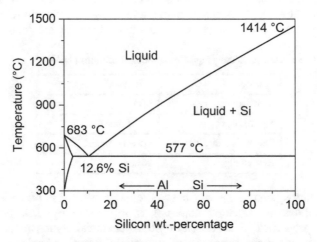

Fig. 2.14:   Aluminum-silicon phase diagram, adapted from [144]

Due to low solubility in Al, Si precipitates are formed as pure silicon. α-phase in the transformation is predominantly aluminum and β predominantly silicon. Depending on the concentration of silicon, the reaction can be classified as hypoeutectic (Si < 12%), eutectic (12-13% Si) or hypereutectic (14-25% Si). Due to low solubility, Si precipitates appear which improve the wear resistance. Size, morphology and distribution of the Si precipitates may vary at different locations of the cast structures which may result in different mechanical properties. Hypereutectic alloys have a wide range of solidification which makes it difficult to cast as well as to be further manufactured due to high Si content. However, hypoeutectic and eutectic alloys are very effective to be cast due to fluidity and machined afterwards. Due to eutectic formation, they have good weldability as well as low solidification shrinkage [144,145].

AlSi binary composition near to the eutectic stage is characterized by acicular or lamellar microstructure consisting of eutectic in aluminum matrix. For hypoeutectic composition, primary Al-dendrites are nucleated at the liquidus temperature. In the solute field, β particles are nucleated slightly above the eutectic temperature. Hypereutectic alloys exhibit more primary Si in a eutectic mixture.

This primary Si is coarsely distributed in the mixture and causes poor mechanical properties [143].

2.4.3 Mechanical properties and applications

Pure aluminum has a very low tensile strength (90 MPa). However, strength of aluminum alloys can be increased by alloying it with other elements as well as heat treatment and can reach up to 700 MPa. This strength combined with its light-weight results in good specific strength which makes it attractive for light-weight design. Fig. 2.15 gives an overview of specific tensile strength of different alloys. Titanium alloys can reach higher specific strength, but their high cost limits their applications to high-end niche markets [142].

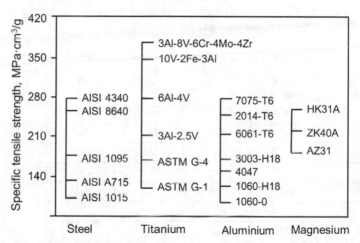

Fig. 2.15: Comparison of specific tensile strength of different alloys, adapted from [142]

Mechanical behavior of aluminum alloys depends on a number of microstructural parameters e.g. coarse intermetallic compounds, dispersoids, precipitates, grain size and morphology and dislocation structure. Coarse intermetallic compounds are formed by eutectic decomposition across the dendrites and are harmful for ductility. Elimination of these compounds incurs high costs. Dispersoids, size ranging from 0.05-0.50 µm, are formed by precipitation in the solid state. They are responsible for hindering the recrystallization and, therefore, grain growth during heat treatment. Fine precipitates (up to 0.1 µm) are nucleated during age hardening and are a favorable source of strengthening. Fine grain size results in higher strength, but usually an accompanying decrease in ductility is obtained.

Dislocation structures are formed during cold working and blocks the respective regions to age hardening [142].

The influence of Si content on mechanical properties is represented in Table 2.2. The increase in strength as a function of Si content increases until eutectic composition, and starts decreasing after that due to formation of coarse Si particles; whereas Si content has a linearly reverse effect on ductility which decreases as the Si content increases. Fatigue strength of the non-ferrous alloys is not necessarily improved by increasing the tensile strength. The fatigue ratio depends on how a material's tensile strength is depending on precipitation hardening (PH). The more a material is dependent on PH for its tensile strength; the lower is its fatigue ratio. Localized straining of precipitates decreases the fatigue strength. Thermo-mechanical treatments increase dislocation density resulting in improved fatigue strength [146].

Table 2.2:   Mechanical properties of AlSi alloys as a function of Si content [146]

Compo-sition Si wt.-%	Tensile strength [MPa]	Yield strength [MPa]	Elongation $[10^{-2}]$	Hardness [HV]	Density $[kg/m^3]$
AlSi2	127.3	52.6	12.4	39.5	2.68
AlSi4	142.2	58.3	10.2	47.3	2.67
AlSi6	155.7	64.8	9.6	55.6	2.65
AlSi8	169.6	71.5	7.2	61.6	2.62
AlSi11.6	185.4	80.0	5.8	67.0	2.59
AlSi12.5	189.0	82.5	5.4	70.0	2.57
AlSi15	183.3	77.7	4.7	72.5	2.55
AlSi17	175.8	73.7	3.0	76.6	2.53
AlSi20	172.4	72.0	2.5	81.0	2.50

Owing to the listed properties, AlSi alloys are used in many moving applications – automotive, aerospace, defense industries for instance. There is a trend to shift engine blocks from cast iron to AlSi alloys to achieve light-weighting benefits. They are used in cylinder heads, pistons, power trains, intake manifolds, chassis, transmission and heat exchangers. Manufacturing of intricate heat exchangers by making use of additive manufacturing is also under development. Manufacturing of many other AlSi concept components by additive manufacturing is envisaged in the next years [143,147].

3 Investigation methodology [1]

To ensure reliability of the structural components in quasistatic and cyclic applications, it is important to understand the part parameters as a function of process parameters and post-processing steps. Understanding of process-structure-property relationships is important for part- and application-oriented processing of SLM parts. Every process parameter and step in the process chain until the end-use effects the part properties and, therefore, functional integrity. This chapter explains the process chain followed for manufacturing of test specimens including process parameters and post-processing steps. It then covers specimen geometries for testing part and mechanical properties as well as the test systems and methodology used for measurement and analysis of results.

3.1 Processing setup and design of experiments

Raw material of the investigated alloy, AlSi12, was obtained from *SLM solutions* in powder form. All the specimens in this study were manufactured in laser beam melting system *SLM 250^{IIL}*. The system is installed with a 400 W fiber laser; the building platform, the base plate, can be heated up to 200 °C. *SLM-AutoFabCAM* software is integrated to the machine system. Argon was used in the building chamber as inert gas. Laser scanning was carried out employing so-called chessboard scanning strategy (Fig. 3.1a). The single islands or chess fields contain hatches at 90° orientation between the fields. The field of orientation was rotated at 79° in each layer (Fig. 3.1b). The specimens were manufactured in different geometries (Fig. 3.2) to be post-machined to final geometries. For surface and microstructural characterization, cubical specimens of 10x10x15 mm^3 were manufactured. Cylindrical specimens of Ø9 mm and length of 110 mm were manufactured for quasistatic and high cycle fatigue tests, and Ø12 mm with length of 90 mm for very high cycle fatigue tests. The effect of energy density, base plate heating and post-process stress-relief was investigated. The effect of energy density is explained in section 2.3 for the output part parameters that a specific energy density is required for an optimal part quality. However, the individual effect of the parameters as laser power, scan speed, hatch distance and layer thickness cannot be ignored for their influence even when the accumulative energy

---

[1] Figures used in this chapter are partly published in [51,129,148–155].

© Springer Fachmedien Wiesbaden GmbH, part of Springer Nature 2019
S. Siddique, *Reliability of Selective Laser Melted AlSi12 Alloy for Quasistatic and Fatigue Applications*, Werkstofftechnische Berichte | Reports of Materials Science and Engineering, https://doi.org/10.1007/978-3-658-23425-6_3

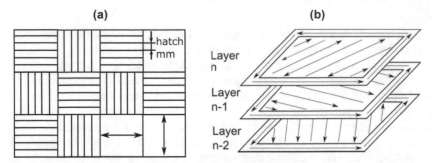

Fig. 3.1:   Chessboard scanning strategy employed for specimen manufacturing (a) [51]; and the rotation of layer orientation in alternating layers (b)

Fig. 3.2:   Exemplary cubical and cylindrical specimens over base plate after SLM processing

density is kept constant. The effect of these processing parameters on, for instance, balling phenomenon is important as lower laser power and large layer thickness will induce large balls; whereas high scanning speed will result in small-sized balls with a higher number. The aim is therefore to use the set of these parameters which not only result in an optimal energy density but which also ensure that the effect of individual parameters on the part properties is taken into account. In this study, values for laser power, scan speed, hatch distance and layer thickness were initially screened for optimal values and then the influence of energy density as well as base plate heating and stress-relief is investigated. The focus lies on the effect of accumulative parameters on part properties such that the extended profile of mechanical properties can be investigated in detail. For this purpose, two levels of each factor were investigated. An experimental plan was developed for the investigation of the stated parameters and given in Table 3.1.

Table 3.1: Experimental plan for optimization of quasistatic properties

Batch	Factors		
	Energy density [J/mm³]	Base plate heating [°C]	Stress-relief [°C]
A	20	0	0
B	20	0	240
C	20	200	0
D	20	200	240
E	39.6	0	0
F	39.6	0	240
G	39.6	200	0
H	39.6	200	240
J	HIP *		

* Hot isostatic pressing

After basic metallographic characterization and quasistatic tests, an experimental plan for investigation of fatigue properties was designed. Fatigue tests were carried out only for the optimal density specimens. Experimental plan for high cycle fatigue (HCF), very high cycle fatigue (VHCF) and fatigue crack propagation tests (da/dN) followed the scheme elaborated in Table 3.2.

Table 3.2: Experimental plan for HCF [1], VHCF [2] and da/dN [3] tests

Batch	BP [4] heating [°C]	Stress-relief [°C]	Additional features	Mechanical testing
E	0	0	-	HCF
F	0	240	-	HCF, VHCF, da/dN
H	200	240	-	HCF, VHCF, da/dN
J	200	240	HIP [5]	HCF
K	200	240	rough	HCF
L	200	240	SP [6]	HCF

[1] High cycle fatigue; [2] Very high cycle fatigue; [3] Rate of crack propagation;
[4] Base plate; [5] Hot isostatic pressing; [6] Shot-peening

3.2   Post-processing

Microstructural and surface observations were carried out using cubical specimens cold-embedded which were ground and polished until a final grit size of 0.3 µm, where oxide polishing suspension (OPS) was used for final polishing. The specimens were observed under optical microscope from *Olympus* for surface analysis of the sectioned specimens where *DHS* software was used for determination of relative density using image correlation. Different sections were observed taking twelve images for each specimen. For microstructural analysis, etching was carried out using HNO_3 and NaOH, and the specimens were sputtered with carbon before taking into scanning electron microscope (SEM) *Tescan Mira XMU (EOS)*. Energy dispersive X-ray spectroscopy (EDX/EDS) analysis was performed inside SEM equipped with *EDAX* detector. The specimens were placed at a distance of 17 mm from the detector. The reflecting signals from the specimens were always larger than 10,000 cps to make the quantification results reliable. This value was achieved by setting the intensity of the applied electron beam to a high value which was 14-16 $W/\mu m^2$ in this case.

Machining was carried out to obtain the specimens in final form for the mechanical tests. Turning was used for machining of specimens for tensile tests, high cycle fatigue tests and very high cycle fatigue tests; whereas milling together with wire erosion was used for the specimens of crack propagation tests. Hot isostatic pressing was carried out on some of the specimens so that the remnant porosity can be completely eliminated. HIP was carried out on the machined specimens at 497 °C and 1000 bar for a holding time of 2 hrs. using argon gas. Shot-peening was performed to investigate the effect of improved surface condition and induction of compressive residual stresses on the fatigue performance. It was performed at a pressure of 2.7 bar at an impingement angle of 90° and a stand-off distance of 50 mm. Average size of the shot was 50 µm. Residual stresses in the specimens were measured using $Sin^2\Psi$ diffraction method using CuK_α-beam with $2\Theta = 137.47°$. The stresses were measured in the gage length of the machined specimens at the surface of the specimens. For the fitting technique, intermediate Lorentz function was applied. Hardness calculations were carried out on software *KB Hardwin XL*.

3.3   Non-destructive defect detection

As the SLM-manufactured parts contain inherent porosity, it is important to characterize this porosity so that it can be analyzed for its effect on mechanical properties. Characterization of such porosity is carried out using conventional destructive 2D surface analysis techniques – optical microscope, scanning electron microscope. However, X-ray-based radiography provides an advantage to measure process-incited porosity in 3D manner. Such non-destructive volumetric measurement is termed as X-ray micro-computed tomography (XCT). One of XCT measurement system is *Nikon XT H 160* consisting of three main units i.e. control unit, X-ray chamber and reconstruction unit. The functional principle, as elaborated in Fig. 3.3, consists of X-ray source and an electrode to emit these X-rays.

Specimen is mounted on a motorized manipulator which can rotate 360° at a small speed so that X-rays can travel through each section of the specimen. These rays are then passed through the projection panel and a projection of the specimen is obtained. *Nikon* X-ray system is equipped with 160 kV / 60 W micro-focus X-ray tube with a focal spot smaller than 200 microns.

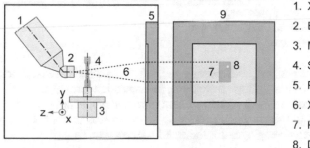

1. X-ray source
2. Electrode
3. Manipulator
4. Specimen
5. Projection panel
6. X-rays
7. Projection
8. Defect
9. Detector

Fig. 3.3:   Functional principle of X-ray micro-computed tomography used for non-destructive defect detection

In order to get an ideal contrast for scanning, calibration of X-ray source filament is required for optimal shading correction which removes the artefacts caused by varying responses between pixels. High resolution amorphous silicon-based *Varian® PaxScan 1313DX* detector is equipped for capturing images. The effective detector area is 13x13 cm^2 with corresponding pixel count of 1024x1024 and 127 μm pixel pitch. The detector is equipped with cesium iodide (CsI) scintillator conversion screen with energy range of 40-160 kVp. The integrated software is capable of capturing images at a rate of 30 frames per second. Specimen is mounted on a motorized manipulator which is capable of 5-axis movement. The linear movement of manipulator is 200 mm in x-axis, 300 mm in y-axis and 610 mm in z-axis.

Three-dimensional representation of the scanned specimen can be achieved by reconstruction of 2D image stack. To create a reconstructed image stack, the system acquires a number of sets of angular views of the specimen called projections. As the XCT system is equipped with a linear detector, the projections are saved as individual lines with an image file, termed as sinogram. The vertical size of the image, i.e. the number of lines in the sinogram images, is the same as the number of projections. Reconstruction of a sinogram creates a reconstructed 2D slice through the specimen. In order to detect the microscopic features of the specimen, X-rays should pass through the specimen. The intensity of transmitted photons passing through the specimen can be described in eq. 3.1 by Beers' law:

$$I = I_0 \cdot e^{-\mu x} \qquad\qquad (3.1)$$

where:

I: Intensity of transmitted photons,
I_0:   Initial intensity of photons,
μ: Linear attenuation coefficient, and
x: Distance travelled.

The quality of the scanned images depends on certain factors such as resolution of scan, brightness and contrast. These factors can be controlled through selection of optimum scan parameters which includes beam energy (kV), beam current (μA) and exposure time (ms). Such parameters can be controlled through CT agent software. The parameters employed for the scans are listed in Table 3.3.

Table 3.3:   Scanning parameters of micro-computed tomography

Beam energy	Beam current	Power	Exposure rate	Effective pixel size	Captured images
130 kV	52 µA	6.8 W	2 fps	7 µm	1,568

Data from computed tomography (CT) was used as input to finite element analysis program, *Abaqus*, to analyze the behavior of process-induced pores under mechanical loading. CT data was converted into STL (Standard Tessellation Language) format and was reconstructed using software *Simpleware*. Software *+CAD* was used to get voxel-based porosity data from pixel-based STL data. Voxel-based data was used to construct a solid model with a voxel size of 15 µm. 4 and 5 voxels were used to construct the solid model such that pores of size larger than 60 µm and 75 µm could be detected, respectively. Gaussian filter is used in this procedure as an algorithm for surface determination. An evolving curve is used which represents a frontier between foreground and background and the intensity between the two is minimized. In the procedure, a so-called island extraction is employed using number of pixels as input and data of connected pixels are termed as islands. The assignment of an island to foreground or background is determined based on the average intensity of the pixels. The resulting model was meshed using tetrahedral elements with a unit size of 6 μm. The meshed model was exported to *Abaqus* where experimental data of mechanical tests was used to simulate the behavior of the material assuming it as non-linear elasto-plastic material. Stress concentration factor (K_t) was calculated for the interesting pores and data from this analysis was used for characterization of fatigue scatter. The transformation is elaborated in Fig. 3.4.

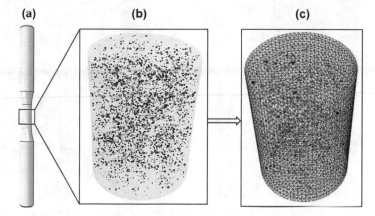

(a)   **(b)**   **(c)**

Fig. 3.4:   Test specimen (a), STL data from computed tomography (b), and meshed model in Abaqus (c) [129,149]

3.4   Quasistatic tensile testing

Quasistatic tensile testing was carried out at the machined specimens according to ISO 6892-1:2009 employing an *Instron 3369* system, overview of which is shown in Fig. 3.5a. It is equipped with a load cell of 50 kN. For calculation of the strain development during the test, an extensometer with a gage length of 10 mm was used (Fig. 3.5b) and the tests were carried out at a strain rate of $1.67 \cdot 10^{-3}$ s^{-1}. The specimen geometry used for tensile tests is shown in Fig. 3.6.

Quasistatic tensile tests were also carried out inside X-ray computed tomography chamber so that CT scans could be performed at intermittent levels in the tensile test for monitoring of the porosity distribution under loading. XT H 160 from Nikon was used for in-situ testing under quasistatic tensile loading inside computed-tomography chamber. In-situ test module *CT5000-GCT-RT* from *Deben UK Ltd.* was used for in-situ tensile testing which is equipped with a load cell of 5 kN. This test module was integrated inside CT-chamber on a platform over the manipulator as shown in Fig. 3.7a.

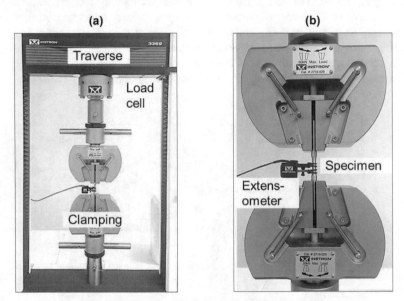

Fig. 3.5:   Instron 3369 system used for tensile testing: overview (a); magnified clamping and gage region (b)

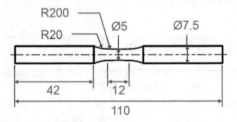

Fig. 3.6:   Specimen geometry for quasistatic tensile tests, load increase tests and constant amplitude tests in the HCF range, dimensions in mm [152]

The test was performed as displacement-controlled at a rate of 1 mm/min as interrupted test. The specimen was scanned before the start of loading (Stage 0), two stages in-between at two different levels of applied force (Stage 1 and Stage 2) followed by a final scan after failure of the specimen (Stage 3), so that the damage behavior during the tensile test could be monitored. A miniature specimen was designed (Fig. 3.7b) which could be clamped in the test system. The data was later evaluated by reconstruction of the two-dimensional images.

(a) **(b)**

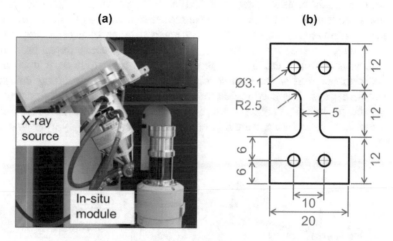

Fig. 3.7: Representation of in-situ tensile module integrated inside X-ray chamber (a); and geometry of the tensile specimen, dimensions in mm, specimen thickness 2 mm (b)

3.5 Fatigue testing in HCF range

Fatigue testing in high cycle fatigue (HCF) range was carried out using two different types of tests - load increase tests (LIT) and constant amplitude tests (CAT). Fatigue tests were carried out only on high energy density specimens due to sensitivity of fatigue loading to porosity. Load increase test was developed by scientists [156,157] which is a part of a short-time test procedure for fatigue testing. During optimization, several configurations need to be tested for fatigue. Fatigue testing using Woehler curves is a tedious process requiring sufficient amount of tests per configuration. Load increase test can be used as initial test for determination of critical stress level for the proceeding constant amplitude tests. The combination of load increase test and a number of constant amplitude tests can, therefore, help in reducing the time in process development. Such procedure has been successfully used for several material classes [80].

A schematic of a load increase test is represented in Fig. 3.8. In such a test, loading is started at a stress level at which no damage in the material is expected; in the current study, it was 30 MPa. The stress is increased at a rate ($d\sigma_a/dN$) such that the test speed is sufficient enough to result in relatively quick results and is slow enough to ensure that the test result is not affected by high loading rate. The loading rate for the tests was kept at 10 MPa/10^4 cycles in a continuous manner. The specimen is equipped with sensor techniques to measure physical parameters like plastic strain amplitude, change in electrical resistance, change in temperature, high frequency impulse measurements, Barkhausen noise etc. depending on the material. Plastic strain amplitude was measured in this study as a measure of material damage during the course of loading. The corresponding course of material damage ($\varepsilon_{a,p}$) is used to determine the stress level until the first material response occurs indicating that the micro-pores formed by dislocation motion have now been developed into a technical crack which grows steeply after that until failure.

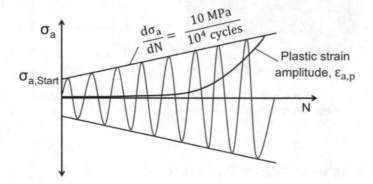

Fig. 3.8:   Schematic of a load increase test with the corresponding course of material damage in the form of $\varepsilon_{a,p}$

The fatigue tests were carried out on an *Instron 8872* (Fig. 3.9a) servohydraulic test system equipped with a load cell of 10 kN. The specimen was clamped in the hydraulic-controlled clamps so that consistent clamping pressure could be applied to avoid any influence of clamping force. Test control of the system is executed by the accompanying software *WaveMatrix™*. An extensometer with a gage length of 10 mm was employed as shown in Fig. 3.9b and tests were conducted at fully reversed cyclic condition, load ratio R = -1, and at a frequency of 20 Hz using test specimen with geometry as shown in Fig. 3.6.

Load increase tests together with constant amplitude tests were used in the process optimization phase which sufficiently reduced the optimization time. Stress level corresponding to non-monotonous development of plastic strain amplitude can be considered as range of critical stress where further tests need to be carried out. A stress level within this range was selected and three constant amplitude tests were performed to determine the mean value and scatter of fatigue life for different batches, so that the best process configuration could be selected. Complete Woehler curve was developed for a set of selected configurations.

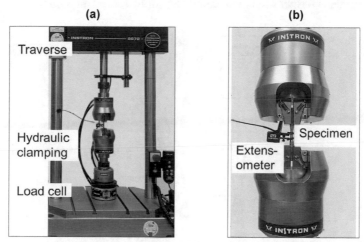

Fig. 3.9:   Overview of the servohydraulic test system *Instron 8872* (a); magnified view of the clamped specimen with installed extensometer for measurement of plastic deformation (b)

3.6   Fatigue testing in VHCF range

Very high cycle fatigue (VHCF) testing was carried out using an ultrasonic fatigue (USF) testing system (*USF-2000A*) by *Shimadzu*. The functional principle of the USF system is elaborated in Fig. 3.10, and the overview of the test equipment is shown in Fig. 3.11a, with calibration plan shown in Fig. 3.11b. Actuator works on the principle of piezo-electricity which resonates at a frequency of 20 kHz when voltage is applied so that only one end of the specimen is clamped, the other remains free. The vibrations are transferred to the specimen through a horn and are designed in a way that resonating longitudinal waves are transmitted through the solid bodies such that the displacement is maximized at the free end of the specimen, and the maximum stress is obtained at the center of the specimen.

Main components of an ultrasonic system include an ultrasound generator to generate ultrasonic waves which are converted to mechanical vibrations by piezo actuator as elaborated in Fig. 3.11a as overview and the calibration setup amplified in Fig. 3.11b. To amplify the amplitude, additional booster and horn are employed to get the maximum displacement amplitude at the free end of the specimen. A data logger is required to formulate a relationship between displacement amplitude at the free end of the specimen, measured by calibrator, and the corresponding voltage which is further used to calculate the applied stress amplitudes.

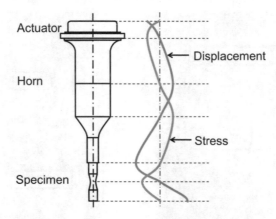

Fig. 3.10:   Functional principle of an ultrasonic fatigue testing system [153]

When high frequencies are used, temperature of the specimen may increase, therefore tests were carried out at intermittent vibrations at a pulse:pause ratio of 50:50. The system experienced resonance for 200 ms and was set to stop for the next 200 ms. Additionally, air cooling was used throughout the tests. As a result, stress- and frequency-dependent increase in temperature was limited to a value below 10 K.In such a system, there is no load cell to measure the stress via force and specimen diameter. Therefore, indirect stress measurements are carried out. Using an eddy current sensor as shown in Fig. 3.11b, the displacement at the lower end is measured. Using the relationship between displacement and strain, based on elastic behavior, value of the strain at the middle of the specimen is calculated. The control variable of the *USF-2000A* is the relative input power of the system. A calibration has to be carried out in order to link the input power with the measured displacement and the stress respectively. As the first step, the change in output signal of the eddy current sensor is correlated to the change in displacement between the sensor and the free end of the specimen. Secondly, dynamic calibration is carried out i.e. a relationship between machine power and amplitude, as measured by the eddy current sensor, is established. Both the steps of calibration were repeated several times to account for statistical errors. Tests were carried out using the specimen geometry shown in Fig. 3.12 which was optimized to resonate at a frequency of 20 kHz.

(a)   **(b)**

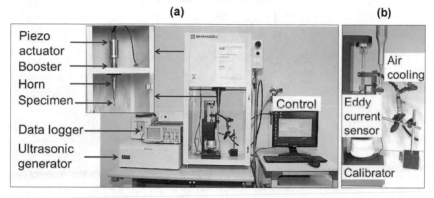

Fig. 3.11:   Main components of ultrasonic fatigue testing system USF-2000A (a); and the magnified calibration setup (b)

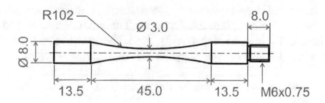

Fig. 3.12:   Specimen geometry for very high cycle fatigue testing, dimensions in mm [153,154]

3.7   Crack propagation testing

Crack propagation tests were carried out to determine the crack growth rate per cycle (da/dN) as a function of stress intensity factor range (ΔK) as well as to find out the threshold stress intensity factor range (ΔK_{th}) and critical stress intensity factor range (ΔK_c). Specimens for crack propagation testing were manufactured in the form of slabs and machined afterwards to final geometry as shown in Fig. 3.13. Crack starting notch was generated by wire erosion using a wire of 0.25 mm diameter. The tests were carried out following ISO 12108:2012 on a servo-hydraulic fatigue testing system *Instron 8801* installed with a load cell of 100 kN. Crack propagation was measured using a crack opening displacement (COD) gage with a gage length of 5 mm and an extension capacity of 2 mm.

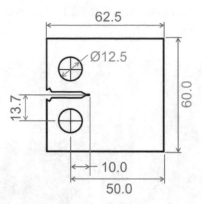

Fig. 3.13: Specimen geometry for crack propagation tests, dimensions in mm, specimen thickness 12.5 mm [153]

To simulate the machined notch as fatigue crack, pre-cracking was carried out at constant ΔK, threshold stress intensity factor range was determined using K-decreasing mode and Paris region was determined at constant force. The control parameters for the three regions are listed in Table 3.4, where a_i and a_f are the initial and final crack lengths respectively, ΔF is the applied force range, and ΔK_i is the initial stress intensity factor range in the respective test region.

Table 3.4: Test control parameters for different regions of fatigue crack propagation test

Pre-cracking	Threshold region	Paris region
Mode: constant ΔK	Mode: K-decreasing	Mode: K-increasing
$\Delta K = 4$ MPa$\sqrt{m}$	K-ratio = - 0.2 mm^{-1}	$\Delta F = 2300$ N
$a_i \approx 10$ mm	$\Delta K_i = 4$ MPa$\sqrt{m}$	$a_i \approx 14$ mm
$a_f = 12$ mm	$a_i = 12$ mm	$a_f = 45$ mm

3.8   Fatigue prediction methodology

A   Fracture mechanics-based approach

Two-fold methodology for fatigue prediction is investigated, first using fracture mechanics approach and the other using plasticity-based approach. The current defect-based approaches predict the fatigue limit of a part; however, keeping in view the existence of fatigue limit being under question, it is required to predict the fatigue life of a part which takes into account the defect data and uses statistical techniques to predict fatigue life at a certain stress amplitude as well as fatigue scatter, the quantification of which is very important for reliable applications. A stochastic approach was therefore developed using Weibull distribution function, maximum likelihood function and crack propagation testing. Fig. 3.14 shows the methodological framework employed for the prediction methodology based on fracture mechanics. Geometrical features determined by micro-computed tomography (μ-CT) analysis were used as input to determine stress raisers, together with constant amplitude load (CAL) fatigue tests and crack propagation (da/dN) tests. Constant amplitude fatigue tests were performed to generate an instance to apply the statistical formulations.

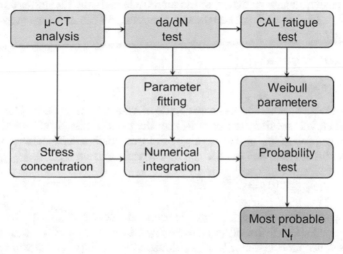

Fig. 3.14:   Framework for fracture mechanics-based fatigue prediction methodology

General formulation of two parameter Weibull distribution is given in eq. 3.2 [95]; based on which, the maximum likelihood function of a two parameter Weibull's distribution was derived and stated in eq. 3.3; thus, deducing the maxima of the natural logarithm of this equation would yield the statistical estimators of the variables of interest α and β as given in eq. 3.4 and 3.5 respectively.

$$f(N_f) = \alpha \cdot \beta \cdot (N_f)^{\beta-1} \cdot e^{-\alpha(N_f)^\beta} \tag{3.2}$$

$$L(N_f) = \alpha^n \cdot \beta^n \cdot \Pi_{i=1}^n (N_f)^{\beta-1} \cdot e^{-\alpha\sum_{i=1}^n (N_{fi})^\beta} \tag{3.3}$$

$$\alpha \approx \sqrt{\frac{n}{\sum_{i=1}^{n}(N_{fi})^{\beta}}} \tag{3.4}$$

$$\beta \approx \frac{1 + \ln(\Pi_{i=1}^{n} N_{fi})}{n \cdot \sum_{i=1}^{n}(N_{fi})^{\beta}} \tag{3.5}$$

Life of a component in the presence of a crack can be deterministically calculated based on the results of crack propagation testing. The rate of crack propagation was determined from the crack propagation tests and the relationship between stress intensity, geometry factor, stress amplitude and crack length are related in eq. 3.6 [91]. The life of a part can then be generalized from the crack growth curve as in eq. 3.7.

$$\Delta K = \Delta\sigma \cdot Y \cdot \sqrt{\pi \cdot a} \tag{3.6}$$

$$dN = \gamma(a)da \tag{3.7}$$

According to the weakest-link approach, life of a part will be the life of the weakest defect which will result in the least life. In the current case, it will be the defect with highest probability to cause fatigue failure, which can be determined using Weibull's distribution, and the life can be determined from eq. 3.8.

$$N_f = \sum_{i=1}^{n} \int_{ai-1}^{ai} \gamma i(a)da \tag{3.8}$$

Weibull parameters α and β were determined according to eq. 3.4 and 3.5 respectively. Defect distribution was obtained from computed tomography, and the stress concentration factor K_t was calculated for each pore. Crack propagation was then numerically integrated for each pore with the calculated stress concentration factor.

B   Plasticity-based approach

In this approach, a representative state of defects was identified on the basis of non-destructive μ-CT and eq. 3.9 [95], which is basically a simplification step towards a more resource-efficient representation of the problem. It depends on the defect size variation which plays an important role in activation of crack initiation. This step is followed by transformation of the μ-CT provided model to the designated modeling environment. In this study, *Abaqus/CAE* was used to apply the direct cyclic analysis shown in the loading spectrum of the simulated cycle shown in Fig. 3.15a, and the corresponding model setup and boundary conditions in Fig. 3.15b. Direct cyclic analysis is beneficial in this case as it uses Fourier transformation together with integration of time of non-linear material behavior such that a stable cyclic response of a structure can be obtained iteratively.

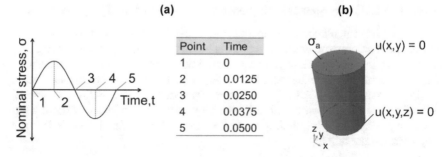

(a)   **(b)**

Point	Time
1	0
2	0.0125
3	0.0250
4	0.0375
5	0.0500

Fig. 3.15:   Abaqus analysis setup: loading spectrum of the simulated cycle with time point translation (a); and model setup and boundary conditions (b)

The load cycle was simulated for a frequency of 20 Hz such that one cycle was completed in 0.0500 s. The boundary conditions were applied such that displacement restriction was applied to the bottom plane in all three directions of translation i.e. x, y and z directions; whereas x and y directions were restricted in the upper plane to get the response in z direction for the applied stress amplitude. The material model is obtained based on the slope per stress amplitude of a continuous load increase test according to eq. 3.10, the aim of which was to obtain a direct cyclic response of the given structure and anticipate cyclic ratcheting and hysteresis relaxation. For realization of the problem, the software relies on Fourier series-based representation of the cyclic displacement in a Newton-Raphson scheme and it's residual. Representation for displacement vector and residual vector can be realized respectively in eq. 3.11 and eq. 3.12.

$$n = [(\frac{z \cdot s}{e})^2]$$
(3.9)

where z is the deviation at a certain confidence interval, s is the standard deviation and e is the expected error.

$$\frac{d\sigma_{a,0}}{d\varepsilon_{a,p}} \approx K(\varepsilon_{a,p})^n$$
(3.10)

$$\overline{U}(t) = u_0 + \sum_{k=1}^{n}[u_k^s \cdot \sin(k\omega t) + u_k^c \cdot \cos(k\omega t)]$$
(3.11)

where n is the number of Fourier terms, and $\omega = 2\pi / T$ the angular frequency.

$$\overline{R}(t) = R_0 + \sum_{k=1}^{n}[R_k^s \cdot \sin(k\omega t) + R_k^c \cdot \cos(k\omega t)] \qquad (3.12)$$

where R_0, R_k^s, R_k^c corresponds to displacement coefficients u_0, u_k^s, u_k^c respectively.

The fatigue life calculation scheme does not depend directly on FEM results, but on the statistical post-processing step. This begins with Dang-Van screening step presented in the framework of plasticity-based approach (Fig. 3.16), where Dang-Van stress limit value can be calculated in eq. 3.13 [96] along with the associated sensitivity factor in eq. 3.14 [158]. Each micro-stress value within the model exceeding this limiting value should pass the screening step into statistical post-processing. Physically implied, any local element stress value below it is not a subject of fatigue failure.

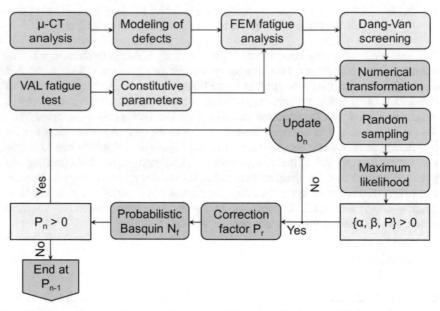

Fig. 3.16:   Framework for plasticity-based fatigue prediction methodology

$$\sigma_{DV} = \tau_{max} + \varphi \cdot P_{max} \qquad (3.13)$$

where σ_{DV} is Dang-Van equivalent stress at any time point, τ_{max} is the maximum shear stress, φ is a sensitivity factor, and P_{max} is the maximum hydrostatic pressure.

$$\varphi = \nu + HV \cdot 10^{-4} \qquad (3.14)$$

where ν is a fitting parameter and HV is Vickers hardness.

The statistical post-processing step begins with construction of the probability density function on the basis of linear numerical transformation according to eq. 3.15. The transformed discrete data is subjected to numerous random sampling, according to eq. 3.9, for aggregation of failure probabilities based on the maximum-likelihood equations of eq. 3.16 and eq. 3.17. Transient damage progression is realized in a Markov-chain of eq. 3.18. Iterations continue until failure probability is maximized which represents the lower limit fatigue life calculated according to eq. 3.19 which is the modified Basquin equation per respective fatigue correction strength for LCF, HCF and VHCF in eq. 3.20, eq. 3.21 and eq. 3.22 respectively. Monte-Carlo curvature (b_m) is used in this case to determine the probabilistic fatigue life N_f^p. At this stage, iteration is not stopped but progressed in Markov-chain until failure probability is minimized, at which the upper bound fatigue life can be obtained.

$$x_i' = b + \frac{x_i - x_{min}}{x_{max} - x_{min}} \cdot (a - b) \tag{3.15}$$

$$\widehat{\alpha} = \frac{n}{\sum_{i=0}^{n} x_i^\beta} \tag{3.16}$$

$$\widehat{\beta} = \frac{n}{1 - \ln\Pi_{i=1}^{n} x_i} \tag{3.17}$$

$$\frac{M}{N} = X_n \cdot \frac{b}{a} \cdot \frac{P_f}{P_s} \cdot \frac{\sigma_{DV}}{K} \tag{3.18}$$

$$N_f^p = \left(\frac{b_n \cdot \sigma_a}{P_r \cdot K}\right)^{1/b_m} \tag{3.19}$$

$$P_r = \frac{P_f}{P_s} \cdot \frac{\sigma_{DV}}{P_{max}} \tag{3.20}$$

$$P_r = \frac{P_f}{P_s} \cdot \frac{\sigma_{DV}}{\sigma_a} \tag{3.21}$$

$$P_r = \frac{P_f}{P_s} \cdot \frac{\tau_{max}}{\sigma_{DV}} \tag{3.22}$$

Correction factor (P_r) was used to employ material strength correction for different applied stresses. The correction factor is used to convert the parameters of load increase test to constant amplitude tests, depending on the different material behavior in different stages of fatigue life and the corresponding applied stress. These empirical relationships are formulated employing stress amplitude σ_a, the corresponding probability of failure and survival (P_f, P_s), and Dang-Van stress amplitude (σ_{DV}). This empirical relationship is also based on the respective failure mechanism in different fatigue ranges, as maximum hydrostatic pressure is more relevant at higher stress amplitudes in the low cycle fatigue range, and maximum shear stress is the dominating mechanism at very small stress amplitudes in the very high cycle fatigue range. Dang-Van stress was more beneficial in prediction of fatigue life in the high cycle fatigue range.

4 Characterization of AlSi12 alloy [2]

4.1 Powder characteristics

Powder material is an important parameter in powder-based additive manufacturing which can affect the quality of the manufactured parts significantly. Therefore, fresh as well as recycled powder material was analyzed under scanning electron microscope to view the size, the morphology and the chemical composition. Fig. 4.1 shows SEM micrographs of the fresh and recycled powder and Fig. 4.2 the corresponding size distribution.

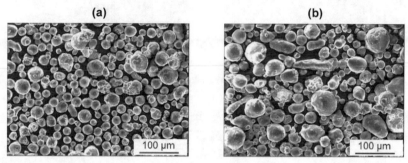

Fig. 4.1: Scanning electron microscope images of: fresh powder (a); recycled powder (b)

Fresh powder is characterized by spherical morphology (Fig. 4.1a) to a large extent with size range of 12-50 µm with a mean size of 23 µm. The distribution around the mean is relatively uniform with a standard deviation of 6 µm (Fig. 4.2a); however, the recycled powder (Fig. 4.1b) shows a mixed spherical and eccentric morphology with a range of equivalent diameter from 8-68 µm (Fig. 4.2b). However, the average size of the recycled powder material (26 µm) is only slightly larger than that of fresh powder.

Chemical composition of the powder material was investigated using energy dispersive X-ray spectroscopy (EDX/EDS) analysis. These investigations were carried out on individual fresh and recycled powder particles using point analysis.

---

[2] Results presented in this chapter are partly published in [10,51,129,153,155,159].

S. Siddique, *Reliability of Selective Laser Melted AlSi12 Alloy for Quasistatic and Fatigue Applications*, Werkstofftechnische Berichte │ Reports of Materials Science and Engineering, https://doi.org/10.1007/978-3-658-23425-6_4

30 measurements were carried out each for fresh and recycled powder particles and their average values are summarized in Table 4.1.

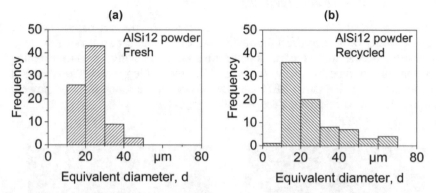

Fig. 4.2:   Distribution of AlSi12 powder size: fresh powder (a); recycled powder (b)

Table 4.1:   Summary of EDX analysis for fresh and recycled powder

Elements	Fresh	Recycled
Al	86.8%	88.6%
Si	12.9%	11.1%

Both the powder batches have Si wt-percentage in a range conforming to the acceptable range of chemical composition for AlSi12 with only a small percentage of minor elements. However, for recycled powder, the percentage of Si decreases from 12.9% to 11.1% which can be attributed to precipitation of Si at the boundaries of the partially melt powder particles. The relatively small difference in the composition as well as size distribution between fresh and recycled powders results due to the fact that the heat affected zone (HAZ) in the SLM process is limited depending on the beam diameter, and that the powder bed is not pre-sintered as in electron beam melting (EBM) process.

4.2   Microstructure

SLM induces a peculiar microstructure (Fig. 4.3) representative of the process nature i.e. very fine microstructure with interacting melt pool boundaries at the laser melt track, accompanied by micro-porosity which is composed of gas porosity. Fig. 4.3a shows the overview of the SLM-induced microstructure of high energy density specimen where no bonding defects are observed; however, small gas porosity do exist. Microstructure consists of parabolic sections which correspond to the grain growth from multiple sides depending on the direction of laser scanning. In totality, it is an Al-Si eutectic in Al matrix which corresponds to the chemical composition in AlSi alloys.

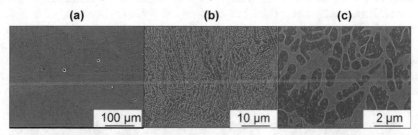

Fig. 4.3:   Microstructure development in AlSi12 alloy (plane parallel to base plate): overview of parabolic melt (a); increasing magnification showing melt track and anisotropic microstructure (b); high magnification image with precipitate formation (c) [51]

Fig. 4.3b shows the microstructural features in detail. Melt track can be seen having coarser structure as compared to that in-between melt tracks. These locations undergo exposure to heat for a longer duration and experience relatively lower cooling rate as compared to internal features. Inside these melt tracks, there is development of microstructure at the local level depending on the cooling rate at specific locations. The cooling rate may vary from location to location depending on the scanning strategy, and for material condition in the case of components. It gives rise to local anisotropy in microstructure. Mostly columnar dendrites grow, the direction of which can differ depending on the direction of heat flow. Higher magnification image of the microstructure in Fig. 4.3c reveals the structure in detail where eutectic grows in the Al-matrix together with the formation of precipitates spread in the matrix.

It is important to recognize that SLM imparts different microstructures in different build directions. Fig. 4.4a is a representative microstructure in xy plane (parallel to base plate) which gives a planar view of the eutectic development in Al matrix; whereas z direction (plane perpendicular to base plate) shows growth of dendrites in the build direction (Fig. 4.4b). A peculiar feature here is the existence of precipitates which are dispersed in the matrix. Another remarkable feature is the spherodization and precipitation phenomena observed at different sections of the SLM-manufactured AlSi microstructure as a function of cooling rate. At the locations where longer exposure is experienced, mainly at the overlap of the melt tracks, cooling rate is relatively low, which results in spherodization of the eutectic phase (Fig. 4.5a). There is a gradual transition from dendritic to spherodized structure in the middle of the transition region as can be seen in Fig. 4.5a where spherodization is observed in the center. Such spherodization of the eutectic phase will result in better ductility as compared to the directional microstructure and can be obtained by controlling the cooling rate in the cases where more toughness is required. However, as the cooling rate changes locally in the process and is higher within the melt tracks (shown in different regions of Fig. 4.3) and at locations near to base plate, other microstructural features are observed at these locations. Si particles are precipitated (Fig. 4.5b) due to the higher cooling rates in the eutectic phase as well as in the matrix. This precipitation can also be a cause of higher strength. mainly

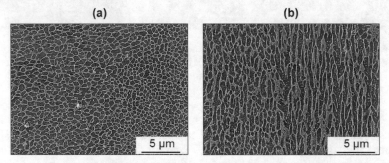

Fig. 4.4:   Microstructural features of SLM-manufactured AlSi eutectic alloy: in xy plane parallel to base plate (a); in z plane perpendicular to base plate (b)

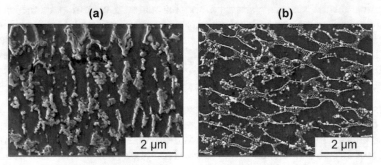

Fig. 4.5:   Characteristic spherodization of the eutectic phase (a); and spherodization of Si particles (b) [159]

Another noteworthy feature regarding the influence of processing and post-processing conditions is the size of dendrites. Width of the dendrites, as developed in the processing, depends on the time available for their growth which is a function of cooling rate. Width of the dendrites was measured across the build direction for the batches E-H. Several images were taken and dendritic width was measured (exemplary measurement in Fig. 4.6) for six images each and the results are summarized in Table 4.2.

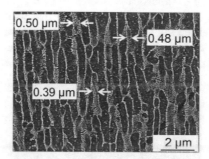

Fig. 4.6: Schematic of measurement of dendritic width from scanning electron microscope

Table 4.2: Dendritic width of the investigated batches E-H

Batch	Base plate heating [°C]	Stress-relief [°C]	Dendritic width [µm]
E	0	0	0.29
F	0	240	0.35
G	200	0	0.51
H	200	240	0.56

Effect of base plate heating as well as stress-relief is an increase in dendritic width. Analysis of variance (ANOVA) suggested that the effect of base plate heating is statistically significant at a significance level of 5%; however, the effect of stress-relief is dominant only in the absence of base plate heating, and is negligible as well as no more statistically-significant when base plate heating is performed. The effect of base plate heating in coarsening of the grains is higher as compared to that of stress-relief. The mechanism behind grain coarsening due to base plate heating is the lower cooling rate when pre-heating is performed. The effect of stress-relief at different levels of base plate heating can be attributed to the potential of recrystallization. Batch E without base plate heating has very fine microstructure i.e. more number of grain boundaries which undergo recrystallization when exposed to heat in post-process stress-relief, thereby coarsening of grains in batch F. However, when base plate heating is already performed (batch G), then the effect of stress-relief (batch H) is not significant, as the microstructure is already coarsened by base plate heating resulting in less number of grain boundaries resulting in very small change in grain size. Such relationship between processing / post-process parameters and the micro-structural features gives guideline to control the microstructure locally by controlling the thermal gradients in the process. Post-process hot isostatic pressing (HIP) treatment recrystallized the microstructure. The treatment at 497 °C and 1000 bar eliminated the eutectic, dendritic micro-structure; and formed a microstructure consisting of Al matrix with Si mixed crystals uniformly distributed in the matrix as shown in Fig. 4.7.

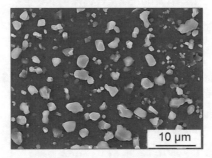

Fig. 4.7:   Microstructure of AlSi12 alloy after hot isostatic pressing (HIP) treatment (batch J) [129]

4.3   Chemical composition

Energy dispersive X-ray spectroscopy (EDX/EDS) was performed on the specimens manufactured with high energy density. As the trace elements remain less than 0.5%, only Al and Si elements were taken into account. Point analysis was carried out in the individual matrix and eutectic phases as shown in Fig. 4.8. Such analysis was carried out on a number of locations in each of the specimens and the average values showed Si content of 5-7% in matrix and 13-17% in the eutectic region. By using rule of mixtures, with eutectic phase ratio of 32-40%, Si content in the overall material was calculated in the range of 10.6-12.3% which corresponds to the eutectic composition of the alloy corresponding to the composition of the powder material. Analysis was carried out on different specimens to investigate the effect of post-process stress-relief and base plate heating on the eutectic phase percentage or element composition. No effect of either of the processes was found significant. It was expected that the Si content would not significantly differ with cooling rates. However, other alloys having solute elements with lower melting and evaporation points can experience element loss by evaporation due to the process nature.

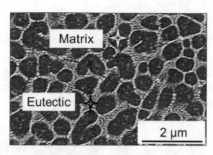

Fig. 4.8:   Example of EDX point analysis in the eutectic and matrix

4.4   Defect morphology

Fig. 4.9 portrays the defect distribution in low energy density batches A and B respectively as observed in optical microscopy images of polished sections. The percentage of porosity in low density batch is calculated to be around 8%. There are bonding defects which occur due to insufficient laser energy required for complete consolidation of powder particles. With this level of porosity, the mechanisms of defect formation are not possible to be comprehended. Fig. 4.10 shows the corresponding surface morphology for high energy density specimens manufactured without, and with base plate heating; with post-process stress-relief, and without that. The images were examined to calculate the relative density. All of these high energy density specimens have relative density higher than 99.5%. The relative density values of the four high energy density specimens is listed in Table 4.3.

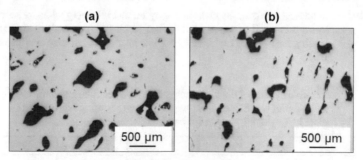

Fig. 4.9:   Surface morphology of polished sections in low energy density batches: as-built batch A (a); and with stress-relief batch B (b)

Batches G and H which are manufactured with base plate heating have higher relative density as compared to batches E and F which are manufactured without base plate heating. This relatively increased relative density is attributed to the higher temperature of base plate which could cause degassing. Base plate heating reduces the thermal gradients which help stabilizing the melt pool. Density of the melt is higher than that of the gases released during melting. When the cooling rate is reduced by pre-heating, it allows more time for gas bubbles to go through the convention process and escape through the melt.

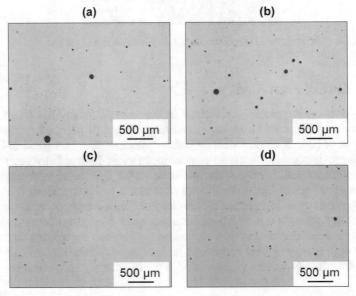

Fig. 4.10:   2D defect detection by metallography in the high energy density specimens: as-built batch E (a); with stress-relief batch F (b); as-built with base plate heating batch G (c); and with base plate heating and stress-relief batch H (d) [153]

Table 4.3:   Relative density of the high energy density specimens: as-built batch E; with stress-relief batch F; as-built with base plate heating batch G; and with base plate heating and stress-relief batch H

Batch	E	F	G	H
Relative density [%]	99.74	99.62	99.87	99.71

Another phenomenon was observed when the relative density of specimens with and without post-process stress-relief is compared: batches E and G (without stress-relief) have relatively higher fraction of relative density as compared to batches F and H (with stress-relief). As the specimens are subjected to higher temperature for extended duration, a pressure develops within the existing pores, which is then responsible for pushing the internal walls of the pores which increases the pore size. Yield strength of AlSi12 alloy drops to 60% at 300 °C [160], therefore considering a closed system of ideal gas containing gas pore with increasing temperature, general gas equation will propose buildup of pressure proportional to temperature increase, which will help soften the alloy resulting in volume expansion of pore. Though the effect is small but was found significant. Such observations are also reported in [161,162].

Non-destructive testing by μ-CT was carried out to observe the volumetric pore fraction. As the selective laser melted parts are subjected to post-process stress-relief to get residual stresses eliminated, μ-CT was performed only to the specimens on which stress-relief was carried out i.e. batches F and H. The volumetric images are shown in Fig. 4.11a and b, and the corresponding defect distribution in Fig. 4.12.

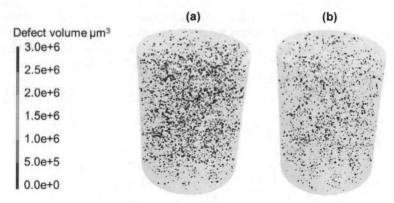

Fig. 4.11: 3D defect detection by micro-computed tomography: as-built with stress-relief batch F (a); and built with base plate heating and stress-relief batch H (b) [153,155]

Density fraction for these specimens from μ-CT was 99.75% and 99.88% respectively. Pore fraction by μ-CT is marginally small as compared to that measured by metallographic image analysis. The trend corresponds to that of metallographic measurements i.e. reduction of porosity by base plate heating. By three-dimensional μ-CT, complete internal structure of the specimen is analyzed, and therefore more reliable as compared to two-dimensional metallography which depends on the cut section. The distribution of defects in Fig. 4.12 shows that after base plate heating, batch H, consistently lower frequency of pores is obtained.

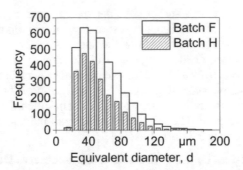

Fig. 4.12:   Stacked histograms for the defect distribution of: as-built with stress-relief
batch F; and built with base plate heating and stress-relief batch H

4.5   Hardness

Fig. 4.13 shows the profile of hardness of the batches E-H as a function of height
from the base plate. The hardness value ranges from 100-120 HV0.2. Among
them, as-built specimen (batch E) has the highest value which can be attributed to
the nature of the manufacturing process i.e. high cooling rates and the
corresponding build-up of residual stresses.

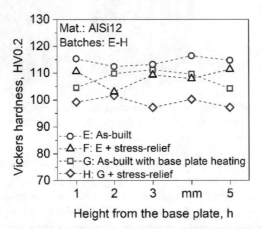

Fig. 4.13:   Hardness profile of the investigated specimens (batches E-H): as-built batch
E; with stress-relief batch F; as-built with base plate heating batch G; and
with base plate heating and stress-relief batch H

After post-process stress-relief treatment at 240 °C (batch F), the hardness value decreased which can be related to the removal of residual stresses. Base plate heating at 200 °C (batches G and H) has resulted in decrease of hardness value. This decrease is representative of the decreased cooling rates as seen in the corresponding microstructure as well as residual stress profiles. The influence of base plate heating as well as post-process stress-relief was found statistically significant at a significance level of 0.05.

The effect of interaction between the two parameters, however, was not found significant as can also be observed from the graphs (batches F and G). Until a height of 5 mm, no correlation between substrate height and hardness value could be observed.

4.6   Residual stresses

Residual stresses are important to be characterized in additive manufacturing processes, especially laser-based processes where high thermal gradients exist causing build-up of thermal residual stresses. These residual stresses can affect the performance of functional applications if these are not measured and taken into account during design phase. These stresses were measured for batches E-H at the surface of the specimens. The results of the surface measurements are listed in Table 4.4.

Table 4.4:   Residual stress profile for the batches E-H: as-built batch E; with stress-relief batch F; as-built with base plate heating batch G; and with base plate heating and stress-relief batch H

Batch	E	F	G	H
Residual stress [MPa]	36 ± 4	13 ± 2	8 ± 2	4 ± 2

The results show that tensile residual stresses are developed in the as-built specimens (batch E) i.e. without stress-relief and without base plate heating, which is about 16.3% of the ultimate tensile strength and 8.6% of the yield strength (section 5.1.1). The magnitude of residual stresses decreases significantly after stress-relief at 240 °C (batch F). These observations are relevant to be discussed on the basis of microstructural features like dendritic width. Fig. 4.14 shows the magnitude of residual stresses as a function of dendritic width. For the as-built batch, maximum residual stress of 36 MPa is measured. The corresponding dendritic width for the as-built batch is 0.29 mm. After stress-relief, the dendritic width increases to 0.35 µm with an accompanying decrease in residual stress to a value of 13 MPa. Effect of the base plate heating (batch G) on relieving the residual

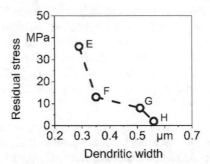

Fig. 4.14:   Residual stresses at the surface of the specimens in the gage length as a function of the dendritic width

stresses is more dominant as compared to post-process stress-relief which is attributed to the decreased cooling rate during the process. The effect is evident also in the increase in dendritic width to 0.51 µm after base plate heating. The interaction effect of base plate heating and post-process stress-relief is the most effective treatment in removal of the residual stresses resulting in about 4 MPa of residual tensile stresses which can be treated as virtual absence of residual stresses.

5 Results and discussions

5.1 Quasistatic and fatigue behavior of AlSi12 alloy [3]

5.1.1 Quasistatic behavior

The quasistatic tensile behavior of SLM-manufactured AlSi12 alloy of the investigated batches A-H (low energy density: A-D; high energy density: E-H) with low and high conditions of base plate heating and post-process stress-relief is portrayed in Fig. 5.1 as characteristic stress-strain curves. Stress-strain curve for batch J, with HIP treatment, is shown separately in Fig. 5.2. Table 5.1 shows the corresponding numerical values together with standard deviation.

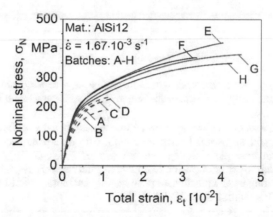

Fig. 5.1:   Exemplary stress-strain curves for quasistatic tensile tests of: Low energy density batches - as-built batch A; with stress-relief batch B; as-built with base plate heating batch C; and with base plate heating and stress-relief batch D. High energy density batches - as-built batch E; with stress-relief batch F; as-built with base plate heating batch G; and with base plate heating and stress-relief batch H [51]

Ultimate tensile strength of the low energy density specimens ranges from 190 to 230 MPa; whereas that of high energy density is in the range of 360 to 420 MPa

---

[3] Results presented in this section are partly published in [51,129,148,150,152–155,159].

S. Siddique, *Reliability of Selective Laser Melted AlSi12 Alloy for Quasistatic and Fatigue Applications*, Werkstofftechnische Berichte | Reports of Materials Science and Engineering, https://doi.org/10.1007/978-3-658-23425-6_5

which implies that the ultimate tensile strength of low energy density specimens is about 55% that of high energy density specimens. For the yield strength, this reduction is about 20%. The fracture strain of low energy density specimens is only $1 \cdot 10^{-2}$ which increased to about $4 \cdot 10^{-2}$ in high energy density specimens.

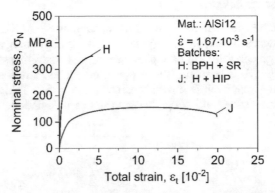

Fig. 5.2: Exemplary stress-strain curves for high energy density batch built with base plate heating and stress-relief batch H; after HIP treatment batch J [129]

Table 5.1: Quasistatic tensile properties of: Low energy density batches - as-built batch A; with stress-relief batch B; as-built with base plate heating batch C; and with base plate heating and stress-relief batch D. High energy density batches - as-built batch E; with stress-relief batch F; as-built with base plate heating batch G; with base plate heating and stress-relief batch H; and with HIP treatment batch J [51,129]

Batch	Quasistatic properties		
	σ_{UTS} [MPa]	$\sigma_{0.2\%}$ [MPa]	$\varepsilon_{t,max}$ [10^{-2}]
A	231.2 ± 5.2	183.9 ± 9.0	1.18 ± 0.09
B	219.5 ± 6.1	180.3 ± 6.5	1.05 ± 0.05
C	190.1 ± 3.4	153.7 ± 4.7	0.98 ± 0.12
D	230.4 ± 3.7	186.5 ± 3.4	1.12 ± 0.10
E	418.9 ± 9.6	220.5 ± 9.4	3.91 ± 0.27
F	372.3 ± 7.2	218.0 ± 6.9	3.41 ± 0.29
G	369.3 ± 3.4	202.2 ± 4.3	4.38 ± 0.16
H	361.1 ± 4.5	201.5 ± 3.7	4.05 ± 0.15
J	155.5 ± 3.4	108.4 ± 2.1	19.20 ± 1.63
Cast [163]	150	70	5

Observing the tensile properties, effect of different processing conditions can be evaluated. Generally, stress-relief has resulted in decreased tensile strength as well as fracture strain (for instance batch E as compared to batch F, and batch G as compared to batch H). The simultaneous reduction in fracture strain and tensile strength is due to the stress-dominated strain behavior during tensile test. The effect of stress-relief is also observed in the specimens manufactured with base plate heating (batches G and H); however, that effect is of less magnitude as compared to that in specimens without base plate heating.

Base plate heating has a particular effect on mechanical properties, as it is expected due to the modified microstructure. Batches G and H, as compared to batches E and F, have higher fracture strain and reduced strength. This effect is attributed to the Hall-Petch effect activated during the process, as the cooling rate decreased due to base plate heating, the microstructure changed into coarser dendritic width (Table 4.2) which is responsible for the modified mechanical properties. However, along with grain boundary strengthening, other mechanisms also come into play. Porosity also has an influence - as the porosity decreases, there is a simultaneous effect on strength as well as ductility which both increase. It can be observed in comparison of batches E and F. Batches E and F are having remnant porosity of 0.26% and 0.38% respectively. An increase in porosity not only decreases the strength but also decreases the fracture strain. The same effect is evident for batches G and H. However, the most important mechanism involved is the Si solubility. Si has very low solubility in Si which is 1.65% at eutectic temperature and reduces to only 0.06% at 200 °C. The Si solubility is increased at high cooling rates. For cast AlSi12 alloy, there is around 10% of free Si, but in SLM-processed alloy it was reduced to 1% [47]. Entrapment of Si in SLM process is the major reason, together with grain refinement, for increased strength. It also explains the differences in strength due to stress-relief and base plate heating. Both of these treatments have resulted in decrease of tensile strength which can be explained by decreased solubility of Si in Al with increased temperature, and due to metastable nature of microstructure, rejection of Si in the matrix. As the Si is rejected from the matrix, it decreases the strength. The accompanying decrease in ductility is also explained based on the free Si, as the free Si is generally brittle and its higher amount results in reduced ductility. Marginally less decrease in fracture strain as compared to a much higher increase in tensile strength, when compared with cast alloy, is explained with free Si and grain refinement. Fine grains generally result in less ductility, however, the reduction of free Si which is usually of cuboidal nature in cast alloy, has resulted in retaining the ductility of the alloy.

The decreased strength and fracture strain in low energy density batches is explained by the fractographic analysis of the tensile test specimens. Fig. 5.3 shows the representative fractographs of batches A and E. Low energy density specimens have a number of bonding defects. Fracture mechanism in low energy density specimens is characterized by crack initiation from the defect regions, and the crack propagation phase is very rapid for these specimens. There are some dimples visible in low energy density specimen (Fig. 5.3a) that means ductile fracture exists but only in the parts where there is no pore. As the number of defects is considerably high, the space between these macro-pores is insufficient to allow for a considerable amount of plastic deformation. Therefore, low energy density batches resulted in decreased fracture strain accompanied by a decreased strength. Fracture surfaces of high energy density specimen (Fig. 5.3b) shows spherical dimples from micro-pores resulting in crack initiation from these pores, which propagated plastically until fracture.

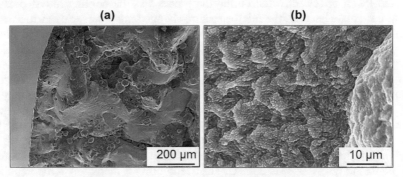

(a)   **(b)**

Fig. 5.3:   Tensile fracture surface of: low energy density specimen batch A (a); and high energy density specimen batch E (b) [51]

After hot isostatic pressing (HIP) at 497 °C and 1,000 bar, stress-strain behavior is characterized by ductile extension with typical necking for ductile materials as shown in the flow curve in Fig. 5.2. The HIP treatment has recrystallized the structure (Fig. 4.7) with coarse microstructure. The tensile strength decreased to about 155 MPa and the yield strength to about 110 MPa with an increase of fracture strain to above $19 \cdot 10^{-2}$ (Table 5.1).

To analyze the effects of the investigated parameters statistically, their main effects are plotted in terms of marginal means in Fig. 5.4. Energy density is of the utmost importance for the tensile properties. However, the effect of energy density on yield strength is considerably small as compared to its effect on tensile strength as well as relatively large difference in imparted energy density and, therefore, the costs.

Base plate heating has consistently decreased the ultimate tensile strength as well as yield strength with an accompanying increase in fracture strain, and the effect is statistically significant (p < 0.05), as was analyzed in multivariate analysis of variance (MANOVA). The effect of stress-relief was not found significant on the quasistatic properties, which is due to the fact that the influence of post-process stress-relief is attenuated when base plate heating is carried out.

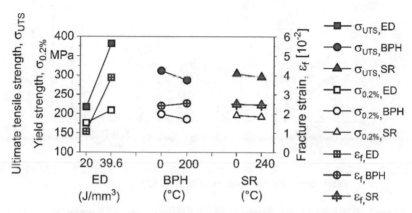

Fig. 5.4: Marginal means of the investigated parameters for quasistatic properties [51], where ED is energy density, BPH is base plate heating and SR is stress-relief

Compared to conventionally-manufactured alloy (Fig. 5.5), ultimate tensile strength of low energy density batches (batches B and D) exceed the value obtained for cast alloy with 50% higher strength in average for low energy density batches. For optimal energy density batches, the ultimate tensile strength for batches F and H is 248% and 240% respectively that of cast alloy. For yield strength, batches F and H result in values 290% and 270% respectively that of cast alloy.

The increase in strength is attributed to the fine microstructure obtained in the process as well as formation of precipitates. Such fine microstructure, however, has decreased the ductility of the SLM parts which is about 12-22% lower than that of cast alloy for optimal energy density batches. The relatively small decrease in fracture strain as compared to high increase in strength values can be attributed to the spherodization of eutectic phase. After HIP treatment, the fracture strain was increased to about 400% with an accompanied reduction in the strength values. The strength values are still comparable to those of cast alloy. For applications where high toughness is required, SLM process is recommended when the geometric complexity hinders its manufacturing by conventional processes, and the structure can be HIPed afterwards to improve toughness.

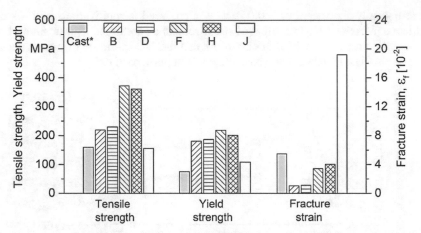

Fig. 5.5:   Comparison of quasistatic properties of selected batches with conventionally-manufactured alloy (* [163])

The small difference in yield strength of the low energy density specimens as compared to high energy density specimens, together with the higher values as compared to cast alloy, recommends using low energy density for applications where structures are not subjected to cyclic loading. The concept can be used for manufacturing the structures with different energy input at different locations of a component with respect to the importance of its respective features. Critical locations can be identified, and the non-critical parts can be manufactured with relatively low energy density to reduce costs.

5.1.2   Damage behavior under in-situ interrupted loading

To determine the damage behavior during quasistatic loading, tensile tests were performed in-situ under X-ray computed tomography where specimens were scanned at different stages during tensile loading. These tests were performed on batch H specimens (manufactured with base plate heating, and stress-relieved) which show rather tolerant behavior under tensile loading due to its relatively less sensitive microstructure and minimal residual stresses. Three tests were performed to ensure the repeatability and to find the statistical values of stress and strain magnitudes. Fig. 5.6 shows an exemplary force-displacement (F-d) diagram as obtained from the in-situ loading module software, and the corresponding stress and strain were calculated according to the test specimen, which is also shown along with.

Specimen was scanned before starting the test (stage 0), two intermediate stages (stage 1 and stage 2), and finally after fracture of the specimen (stage 3). The general behavior of the stress-strain curve shows the phenomenon of stress relaxation at the interrupted stages, which has increased the total strain of the material significantly.

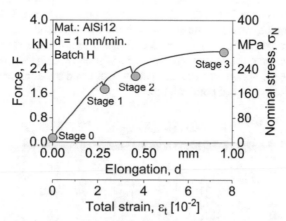

Fig. 5.6: Exemplary force-elongation and stress-strain curve for interrupted tensile test for batch H performed inside computed tomography chamber

The damage behavior was investigated at the interrupted stages and the corresponding images from CT scanning from stage 0 to stage 3 are represented in Fig. 5.7. The statistical calculations for number of defects and defect volume at each stage was calculated and shown in Fig. 5.8.

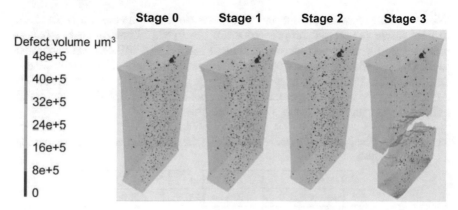

Fig. 5.7: Images from computed tomography for interrupted tensile test from stage 0 to stage 3

Before loading, the scanned section had 911 pores, and mean pore volume of 44,599 μm^3 (Fig. 5.8). After first interruption at stage 1, number of pores increased to 1,048; however, the corresponding mean pore volume increased only marginally to 45,013 μm^3, which shows that though new pores are formed; their size is sufficiently small. At stage 2, number of pores decreased from 1,048 to 1,015 with a small increase in mean pore volume up to 45,268 μm^3, which can be attributed to the start of pore coalescence. Stage 3 results in 390 pores, and the mean defect volume increased up to 62,638 μm^3, which is attributed to a significant increase of pore coalescence which decreased the number of pores with a naturally increasing mean pore volume.

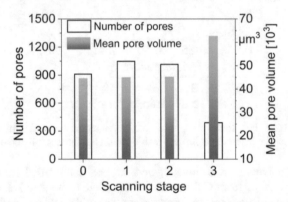

Fig. 5.8:   Representation of pore characteristics for interrupted tensile test in terms of number of pores and mean pore volume at the interrupted stages

When compared with the tensile tests performed continuously (Table 5.2), differences in the quasistatic properties are observed. Ultimate tensile strength obtained from continuously performed tensile test is 361.1 MPa; which decreased to 324.1 MPa when tests were performed in interrupted mode. Correspondingly, the maximum total strain at fracture increased from $4.05 \cdot 10^{-2}$ to $6.36 \cdot 10^{-2}$. Additionally, the scatter in the quasistatic properties in interrupted tests increased as compared to that in continuously performed tests.

Table 5.2:   Comparison of quasistatic properties for interrupted and continuous tensile tests

Test mode	σ_{UTS} [MPa]	$\varepsilon_{t,max}$ [10^{-2}]
Interrupted	324.1 ± 23.1	6.36 ± 1.29
Continuous	361.1 ± 4.5	4.05 ± 0.15

Increase in fracture strain and decrease in tensile strength in interrupted tests can be attributed to the stress relaxation phenomenon (Fig. 5.6). Stress relaxation occurred at the two interrupted stages i.e. stage 1 and stage 2. The magnitude of stress relaxation increased at stage 2 as compared to stage 1 which implies that higher stress relaxation occurred at higher strain level. Such behavior can be explained on the basis of dislocation motion. Increased plastic deformation means higher speed of dislocation motion which can be held responsible for increased stress relaxation, which was, in turn, responsible for increased fracture strain. Such behavior was recently observed by Li et al. [164] for stainless steel 304 where fracture strain increased up to 15% in interrupted tensile tests as compared to continuously performed tests. In the current case, there is an average increase of 57% in fracture strain. Stress relaxation resulting in improved ductility is favorable for damage tolerant designs improving service performance.

5.1.3 Fatigue behavior in HCF range

Fatigue behavior in high cycle fatigue (HCF) range was characterized by a combination of load increase tests and constant amplitude fatigue tests. Fatigue behavior was determined only for the optimal energy density specimens, as the effect of porosity under fatigue loading is well-known to deteriorate fatigue performance. Continuous load increase tests were carried out equipped with measurement of plastic strain to evaluate the course of material reaction during the cyclic loading.

Fig. 5.9 shows that the measurement of plastic strain amplitude portraying the course of material reaction such that the sensitivity of material to applied stress amplitude can be deduced. The material response may be divided into three regions. The plastic strain amplitude remains almost unchanged until about 80 MPa for the three batches, after which it increases slightly due to the formation of micro-cracks until about 110 MPa. After that, the response parameter increases at exponential rate until failure. All the tested specimens in the batches E, F and H failed between 155-175 MPa of stress amplitude. Therefore, the stress amplitude after 110 MPa is critical for these batches to carry on further constant amplitude tests.

Constant amplitude tests carried out at stress amplitudes above this critical value and below the stress amplitude at failure can result in a combined procedure for investigation of fatigue strength in an efficient way such that time- and cost-extensive Woehler curves for the whole range of stress profile need not to be realized for each set of processing conditions, which is relevant specially in the process optimization phase.

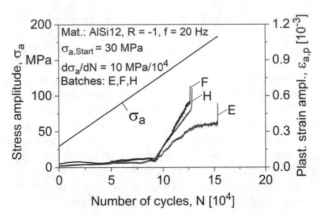

Fig. 5.9: Exemplary course of material response in load increase tests for: as-built batch E; with stress-relief batch F, and built with base plate heating and stress-relief batch H

Load increase test procedure was also carried out for batch J (hot isostatically pressed), batch K (batch H without post-build machining) and batch L (shot-peened specimens) and the results are plotted in Fig. 5.10. Hot isostatically pressed specimen (batch J) failed only after reaching 70 MPa of stress amplitude, rough specimen (batch K) endured until 130 MPa and shot-peened specimen reached to 160 MPa until failure.

The considerably reduced fracture stress in HIP specimen is due to the specific microstructure already observed in section 4.2 which caused higher plastic strains as compared to other batches at comparable stress amplitude (plastic strain amplitude exceeded $0.8 \cdot 10^{-3}$). Such plastic deformation was also observed in their quasistatic behavior (section 5.1.1). To further characterize the fatigue behavior, batch J was not considered due to its lower fracture stress in load increase tests. Stress amplitude of 120 MPa was, therefore, chosen for the proceeding constant amplitude tests.

Fig. 5.11 shows the cyclic deformation behavior in terms of plastic strain amplitude at 120 MPa for different batches of AlSi12 alloy with differences in in-process and post-process configurations. Most important process-specific configurations are represented in Fig. 5.11a in terms of batches E, F and H. Batch E shows the highest life of the three batches, conforming to the result of load increase tests in Fig. 5.9. The cyclic deformation of batch E is characterized by the smallest plastic strain amplitude. All the batches show an initial increase in plastic strain amplitude followed by a decrease i.e. the specimens experience hardening until it is relatively stabilized after 30-60 cycles for different batches. After that, all the batches show different behavior during their remaining fatigue life.

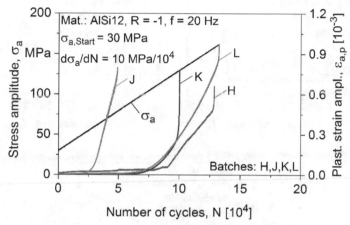

Fig. 5.10: Exemplary course of material response in load increase tests for: hot isostatic pressed batch J, rough batch K, and shot-peened batch L, together with batch H

Batch E, with the as-built condition without any in-process heating and no post-process stress-relief, is characterized by very small changes, after a second stage of small cyclic softening, in the cyclic behavior after initial cyclic hardening until fracture. These small changes, together with the least magnitude of plastic strain amplitude, is representative of the processing route through which it was manufactured. The SLM process, having high cooling rates, imparts very fine microstructure characterized by dendritic width presented in Table 4.2. Such a fine microstructure allows only a small movement of dislocations by offering obstacles by rapid changes in microstructure morphology. Fig. 5.11c shows the influence of dendritic width on the development of plastic strain amplitude at fracture in constant amplitude tests at 120 MPa of stress amplitude for batches E, F and H. As the dendritic width increases due to the post-process stress-relief treatment (batch F), or decrease in cooling rate due to base plate heating (batch H), the plastic strain amplitude at fracture also shows a corresponding increase in its magnitude. Here it is observable that the marginal effect of base plate heating exceeds the effect of post-process stress-relief. The small magnitude in as-built specimens can result in better fatigue resistance; however, the increased magnitude after base plate heating (batch H) shows its increased capacity for a damage-tolerant fatigue behavior.

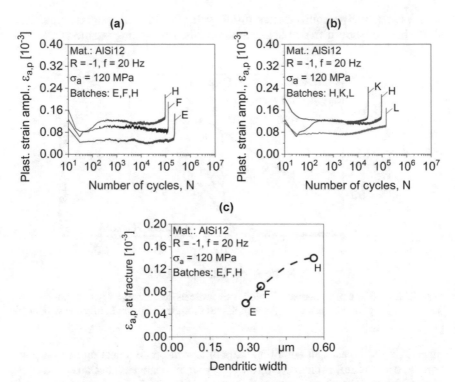

Fig. 5.11: Cyclic deformation behavior in terms of plastic strain amplitude ($\varepsilon_{a,p}$) at 120 MPa of as-built batch E, with stress-relief batch F, and with base plate heating and stress-relief batch H (a); rough batch K, and shot-peened batch L, together with batch H (b); the magnitude of plastic strain amplitude at fracture as a function of dendritic width (c)

Microstructure-based explanation of plastic deformation behavior is relevant for the batches F and H as well. Batch F experiences observable cyclic softening in the second stage, higher than that in batch E, followed by a decrease in plastic strain amplitude in the last stage of fatigue life. Batch H shows cyclic softening higher than batch F, which stabilizes at an intermediate level, and then continues to soften again until fracture. Higher magnitudes of cyclic softening in batches F and H can be attributed to their microstructural features. Batch F has a dendritic width of 0.35 μm and batch H has 0.56 μm (Table 4.2). Based on the explanation of relatively more possibility of dislocation movements in higher grain sizes, differences in magnitudes of plastic strain amplitudes are justified. The differences in deformation behavior in the last stage of fatigue life where a mixture of cyclic softening and cyclic hardening is observed, is explained based on the interaction of microstructural compliance with process-induced defects. Finer microstructures (batches E and F) with process-induced defects result in less possibility of cyclic softening; whereas in relatively compliant microstructure, as observed in batch H, cyclic softening occurred even in the presence of defects.

In the Fig. 5.11b, cyclic deformation behavior for the batches K and L are represented. Both are manufactured with the processing conditions as for batch H; in batch K, no post-process machining was carried out, and post-process shot-peening process was carried out on batch L. Batch K represents similar behavior as other batches at the start i.e. initial cyclic hardening followed by a stable response until cyclic softening is observed before fracture. The magnitude of plastic strain amplitude is similar as in batch H; however, it fractured much earlier due to rough surface. In batch L, also initial hardening is observed, but in the second stage, no cyclic softening occurred as compared to batch H. This behavior is attributed to the formation of a hard layer on the surface of the specimen due to shot-peening where surface compressive stresses were induced, which inhibited plastic deformation at the start; however, in the third stage, this effect vanished and the specimen experiences cyclic softening until fracture.

Cyclic deformation behavior in individual tests of batches F and H at 120 MPa stress amplitude are shown in Fig. 5.12. Batch F has two types of deformation curves. For test 1, it is very small value of plastic strain amplitude which remains almost constant during the course of the fatigue life, with a small portion of cyclic hardening at the start and cyclic softening near the fracture. Test 2 and test 3 have similar cyclic deformation curves with a small difference in magnitude of plastic strain amplitude. Such differences are investigated with the help of fractography and are attributed to the local conditions in the material state regarding process-induced defects. Localized crack initiation near the presence of pores hinder the development of plastic deformation of the material, as was the case in test 1 of batch F (Fig. 5.12a) where multiple crack initiation occurred. The interaction of pores then limits the development of plastic deformation. The other tests (test 2 and test 3) experienced surface crack initiation in the absence of critical localized defects which caused development of plastic deformation to the limit allowed by microstructural features. For batch H (Fig. 5.12b), reduced defects as well as compliance of microstructure even in the presence of small porosity resulted in relatively consistent cyclic deformation behavior for the number of instances tested. These are further explained in fractographic analysis in the subsequent section.

The results of constant amplitude tests at 120 MPa for the investigated batches in terms of average fatigue life (μ) and standard deviation (s) are plotted in Fig. 5.13. Batch E, built without base plate heating and no post-process stress-relief, resulted in highest fatigue life of the investigated batches; however, a considerable fatigue scatter was also obtained for this batch. After performing post-process stress-relief (batch F), average fatigue life as well as fatigue scatter decreased. Batch H, built with base plate heating, has slightly lower fatigue life, but the accompanying fatigue scatter is quite low.

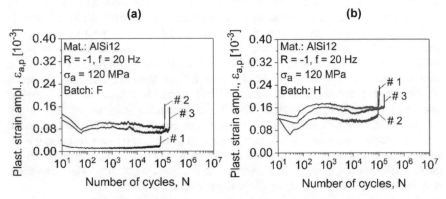

Fig. 5.12: Cyclic deformation behavior in terms of plastic strain amplitude ($\varepsilon_{a,p}$) at 120 MPa in different tests of as-built with stress-relief batch F (a); and built with base plate heating and stress-relief batch H (b)

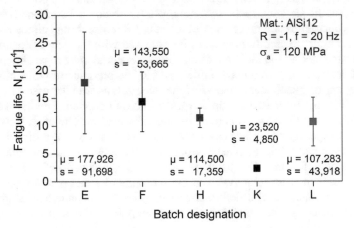

Fig. 5.13: Constant amplitude fatigue tests at 120 MPa resulting average fatigue life (μ) and fatigue scatter (s) for: as-built batch E, with stress-relief batch F, with base plate heating and stress-relief batch H, rough batch K, and shot-peened batch L

The analysis of variance (ANOVA) resulted in insignificant differences in the average fatigue life at 0.05 level, which is due to the high scatter within the batches E and F. The effect of stress-relief on the fatigue life for the specimens without base plate heating is considerable (batch F as compared to batch E). Residual stresses are the potential cause of this effect, as there is relatively small (and statistically insignificant as well) difference which can explain the effect of stress-relief for the specimens built with base plate heating. Stress-induced cracks are produced due to thermal gradients which induced residual stresses. These stress-induced cracks can be a dominant cause of fatigue scatter, as these defects, when become critical, can cause crack initiation from the core of the material. Base plate heating is already shown to decrease residual stresses which reduce the potential of stress-caused cracking, therefore reducing the fatigue scatter.

The trends of fatigue strength and fatigue scatter are to be correlated to the process-induced material parameters. Fatigue strength is related to the microstructure of the specimens, and the effect is a combination of mechanisms in crack initiation and crack propagation phases. In the case of fine grains, crack initiation occurs only from slip bands; whereas in coarse grains, grain boundaries may also be a cause [165]. Material having fine grains will have more grain boundaries which will inhibit the fatigue crack. The effect of grain size on fatigue strength has been found dependent on the applied stress. The effect of grain size is higher at low stress levels where high life is expected. For the stage of crack propagation, fine grain sizes reduce the toughness and result in higher crack growth rates. The total fatigue life is the summation of the two phases, and the favorable grain structure will depend on the dominant phase, crack initiation or propagation, depending on the application.

High cycle fatigue represents a case where fatigue strength is governed by resistance to crack initiation as compared to crack propagation. Resistance to crack initiation is governed by the size of microstructure and, therefore, fine grain size should result in higher fatigue lives in that region. Considering these observations, as-built batch without stress-relief (batch E) has the highest fatigue life of the four investigated batches in Fig. 5.13 which can be attributed to the finest grain size with dendritic width (Table 4.2) of 0.29 μm, followed by batch F with a dendritic width of 0.35 μm obtained by post-process stress-relief. The fatigue life is further decreased in batch H, with base plate heating having dendritic width of 0.56 μm which is relevant from microstructural point of view where dendritic width is sufficiently higher than those without base plate heating, therefore coarser microstructure which is responsible for lower life in high cycle fatigue range.

Another important feature of fatigue which can be recognized from Fig. 5.13 is the fatigue scatter. Batches E and F experience sufficient scatter in decreasing order as compared to batch H. To analyze the cause of failure which could be correlated to the fatigue scatter, fractured specimens were investigated under scanning electron microscope. Representative fracture surfaces are shown in Fig. 5.14 to Fig. 5.16 for these batches.

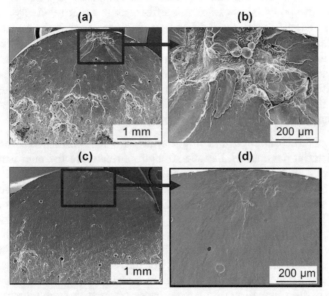

Fig. 5.14:   Crack initiation in as-built batch E: specimen E2 from pore (a), the corresponding crack initiation site magnified (b) [51]; specimen E3 from sub-surface (c), and the corresponding crack initiation site magnified (d)

In the specimens built without base plate heating, there are multiple causes of failure – either from surface or from inside the core from region where bonding was not perfect. Specimen E2 (Fig. 5.14), for instance, experienced crack initiation from a slightly sub-surface location where bonding defect of about 500 μm did exist in the fractured specimen. Such specimens failed much earlier; whereas some specimens experienced fatigue crack initiation from the surface of the specimen, E3 for instance, enduring longer fatigue lives. Similarly, batch F (Fig. 5.15), with stress-relief, shows multiple crack initiation in one of the specimens (F1) and crack initiation from a smaller surface defect in another specimen (F3). Such differences in the failure mechanisms within a batch led to sufficient fatigue scatter. Such inconsistencies were not observed in batch H, with base plate heating, where surface crack initiation has been found as the dominant cause of fatigue failure (Fig. 5.16) which resulted in less fatigue scatter.

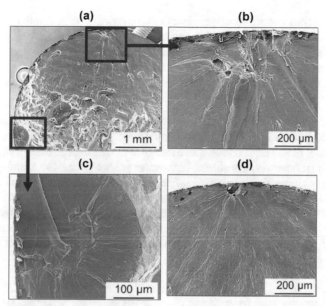

Fig. 5.15: Crack initiation in as-built with stress-relief batch F: specimen F1 from multiple defects (a), the corresponding crack initiation sites magnified (b, c) [51]; specimen F3 from sub-surface defect (d)

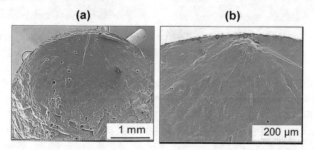

Fig. 5.16: Surface crack initiation sites in specimens built with base plate heating and stress-relief batch H [51]

Batch K (rough) has the least fatigue strength accompanied with very small standard deviation. The small fatigue strength is due to the high surface roughness of the as-built specimen which is average roughness (Ra) of 8 μm and maximum roughness (Rz) of 54 μm. The roughness acts as preferential site of fatigue crack initiation and the existing roughness having an average profile depth of 54 μm act as a notch, and fatigue crack usuallystarts from that notch, therefore, decreasing the fatigue life drastically. The roughness is also attributed to the small standard deviation, as the fatigue crack always started from surface and being the dominant mechanism of fatigue failure. Fracture surface of rough specimen shows the crack initiation from rough surface (Fig. 5.17a).

Shot-peening would have expected to result in a higher fatigue life as compared to that of batch H due to inhibition of crack initiation by compressive residual stresses on the surface; however, it resulted in average fatigue life at 120 MPa slightly less that of batch H; however, the scatter in the case of shot-peened specimen, batch L, is much higher than that of batch H. Fracture surface of shot-peened specimen (Fig. 5.17b) shows that the effect of shot-peening layer can be identified at the boundaries. As it introduces compressive residual stresses at the surface layers, these must be compensated by tensile residual stresses inside the material core. These tensile residual stresses together with the interaction of residual porosity have caused the fatigue scatter

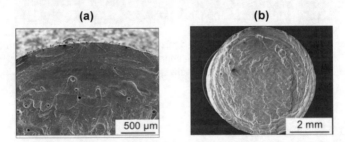

Fig. 5.17:   Fracture surfaces of: surface crack initiation in rough specimen batch K (a); and internal crack initiation in shot-peened specimen batch L (b)

Post-process stress-relief must be carried out on SLM specimens to avoid the high scatter in fatigue. Roughness of the as-built specimens is high enough for their application in cyclic loading. Therefore, post-processing needs to be carried out before they are further loaded. Shot-peening has resulted decreased fatigue strength and increased fatigue scatter due to the interaction of tensile residual stresses in the core of the material with the process-induced defects. Therefore, shot-peening is not recommended as far as there are critical pores existing. Based on these observations, only the batches F and H were considered for further investigations. Fig. 5.18 shows the cyclic deformation behavior of these two batches at two different stress amplitudes. It shows that the increase in plastic strain amplitude by increasing the stress amplitude from 120 MPa to 140 MPa in batch F specimens (without base plate heating) is very low as compared to that in batch H (with base plate heating), which shows that the microstructure of the specimens built without base plate heating has a small margin for plastic deformation, and therefore, their application in fatigue loading especially at highly loaded configurations is not recommended due to their very small compliance. It further shows that the influence of in-process base plate heating has much higher effect as compared to post-process stress-relief when it is loaded under fatigue.

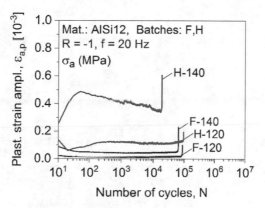

Fig. 5.18:   Cyclic deformation behavior of batches F and H in terms of plastic strain amplitude ($\varepsilon_{a,p}$) under constant amplitude loading at stress levels of 120 MPa and 140 MPa

5.1.4 *Fatigue behavior in VHCF range*

Batches F and H were selected for testing for the extended range of stress amplitudes and the results are portrayed in Fig. 5.19 as complete Woehler curves including high cycle fatigue (HCF) as well as very high cycle fatigue (VHCF) ranges. At 1E9 cycles, the fatigue strength is higher for the specimens built with base plate heating (batch H) as compared to that without base plate heating. However, the trend is not consistent; batch F has fatigue strength higher than that of batch H at higher stress amplitudes; whereas the trend is altered at around stress amplitude of 120 MPa. Such behavior represents the effect of interaction of microstructure and process induced defects at different stress levels. As in HCF, the fatigue strength of a material is governed by resistance to crack initiation which is more in fine grains as compared to coarser ones. Higher life for the specimen at 140 MPa without base plate heating, having less dendritic width, can be comprehended on microstructural basis.

The other important factor is the difference in residual porosity of the two batches. These small pores remain relatively inactive at lower stresses implying that the fatigue life of a specimen in crack propagation phase for the coarser microstructure is higher in proportion as compared to that with finer microstructure. The results suggest that the effect of base plate heating on fatigue life depends on the applied stress amplitude as well.

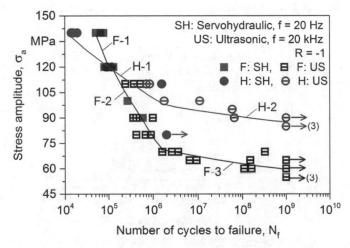

Fig. 5.19:   Woehler curves for AlSi12 alloy: without base plate heating batch F, and with base plate heating batch H until very high cycle fatigue range

Mathematical formulations for the fatigue life (N_f) of the two batches in different ranges are given in eq. 5.1 to 5.5 as a function of applied stress amplitude. According to the fatigue data presented in Fig. 5.19, the corresponding fitting parameters i.e. fatigue strength coefficient (K) and fatigue strength exponent (n') given in eq. 5.1 to 5.5 are of considerable significance. For batch F without base plate heating, three different ranges were identified based on best-fit conditions; whereas batch H with base plate heating could be fitted with two expressions; which implies that batch F has two transition points i.e. one between low cycle fatigue and high cycle fatigue, and the other between high cycle fatigue and very high cycle fatigue ranges. For batch H, only one transition point was obtained which is between high cycle fatigue and very high cycle fatigue ranges. The slopes of lines until HCF range for batch F are much steeper as compared to that in batch H (F1:H1 = 2.46; F2:H1 = 1.97); however, in the very high cycle fatigue range, this difference decreases (F2:H2 = 1.87). The overall slope factor between HCF and VHCF ranges are 2.82 (F2:F3) and 2.67 (H1:H2) respectively for batches F and H, which shows that the effect of fatigue ranges between the two batches is very small, as compared to the with-in batch differences as discussed above as susceptibility to steeper curves in batch F.

$$N_f = \left(\frac{\sigma_a}{1296}\right)^{-\frac{1}{0.204}} \qquad \text{range F-1} \tag{5.1}$$

$$N_f = \left(\frac{\sigma_a}{794}\right)^{-\frac{1}{0.164}} \qquad \text{range F-2} \tag{5.2}$$

$$N_f = \left(\frac{\sigma_a}{188}\right)^{-\frac{1}{0.058}} \qquad \text{range F-3} \tag{5.3}$$

$$N_f = \left(\frac{\sigma_a}{312}\right)^{-\frac{1}{0.083}} \qquad \text{range H-1} \tag{5.4}$$

$$N_f = \left(\frac{\sigma_a}{165}\right)^{-\frac{1}{0.031}} \qquad \text{range H-2} \tag{5.5}$$

At 1E9 cycles, batch F has fatigue strength of 60.5 ± 4.7 MPa; whereas batch H has the corresponding value of 88.6 ± 3.3 MPa. The fatigue strength at 1E9 cycles for the base plate-heated batch is about 45% higher as compared to that without base plate heating. The standard deviation for the pre-heated batch is also lower than that without pre-heating.

Lower fatigue strength accompanied with higher fatigue scatter is attributed to the role of internal defects. Uneven distribution of defects with respect to their location and size is the cause of this scatter, as determined at higher stress amplitude as well. This is exemplified by the fractography of the four specimens at 70 MPa which is representative of the extreme scatter.

Specimens of the tests corresponding to Fig. 5.20a, b and c underwent internal crack initiation from material defects caused by bonding deficiency resulting in reduced life. The relatively bigger size of the defect where crack started from causes a rapid growth of crack. In the presence of several nearby defects of sufficient size, multi crack initiation causes decrease in fatigue life as in Fig. 5.20c. At the same stress level, one specimen endured sufficiently high (Fig. 5.20d) where fatigue crack initiated at a smaller defect resulting in higher fatigue life.

As observed in section 5.1.3, base plate-heated specimens resulted in more reliable fatigue behavior which is due to dominant failure mechanism from surface, as it had less pore fraction giving higher probability of failure at internal pores. Some small fraction of crack initiation from material defects was also observed in pre-heated specimens. Representative surfaces of fatigue crack initiation in batch H are shown in Fig. 5.21.

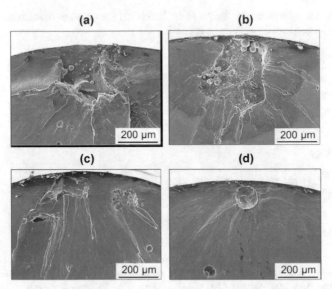

Fig. 5.20: Fatigue crack initiation at 70 MPa for batch F without base plate heating with resulting fatigue life: 1.6E6 cycles (a); 2.6E6 cycles (b); 3.6E6 cycles (c) [154]; and 3.2E8 cycles (d) [153]

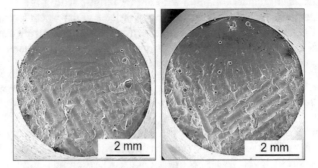

Fig. 5.21: Dominating mechanism of fatigue crack initiation in batch H specimens, built with base plate heating [148,153]

From the existing literature, fatigue strength of cast AlSi alloys range from 60 to 90 MPa for different material and manufacturing conditions from 1E6 to 1E8 cycles. AlSi12 alloy in pressure die cast conditions showed a fatigue strength of 70 MPa to 120 MPa at 2E6 cycles under bending load [166]. In surface layer melted with electron beam melting process, fatigue testing at ultrasonic frequency showed horizontal asymptote in fatigue strength of about 70 MPa at 5E7 cycles [167]. From the additively-manufactured AlSi alloys, fatigue investigations rarely exist; however, for AlSi10Mg alloy, Brandl et al. [44] showed a fatigue strength, in terms of maximum strength at stress ratio of 0.1, of 120 MPa, for as-built and up to 210 MPa at 2E6 cycles after T6 treatment. The corresponding fatigue strengths in terms of stress amplitude are 60 MPa and 105 MPa respectively. The same alloy resulted in fatigue strength of about 90 MPa at 1E7 cycles in rotating bending fatigue [168]. AlSi12 alloy in the current study exhibited a fatigue strength of 67.4 and 92.3 MPa at 1E9 cycles respectively for the configuration built without base plate heating and with base plate heating.

Frequency effect

An important aspect to be discussed is the effect of frequency in the fatigue test results carried out employing servohydraulic test system at 20 Hz and those performed using ultrasonic fatigue testing system at 20 kHz. The topic is of high interest in the fatigue community and several studies have been carried out in the last years to investigate the influences, as the frequency in real applications is usually much lower than ultrasonic frequency. These results need to be compared not only for frequency effects but also due to the testing techniques, as the testing technique may vary depending on the control parameters as well as size effect when fatigue test specimen is considered [169]. Effect of frequency in the case of carbon steels have been confirmed by several researchers [170,171]. They report increase in fatigue strength in carbon steels when tests are performed at ultrasonic frequencies. Though the real reason is still unknown, it is expected that the increase in strength can be attributed to the strain rate effect, where cyclic plasticity is reported to decrease at high frequencies. However, there is still no agreement in the results of different studies. Furuya et al. [172] observed no difference in the fatigue strength of ultrafine grained steel at 150 Hz and 20 kHz. Similarly, no effect of frequency was found in austenitic steels at frequencies of up to 200 Hz and 20 kHz. Also no frequency effect was observed in the study of Furuya et al. [173] on high strength steel.

Different studies investigating frequency effect are also available for aluminum alloys. Mayer et al. [174,175] have investigated Al-Zn alloy at frequency values of 100 Hz and 20 kHz in very high cycle fatigue at fully-reversed as well as pulsating load ratios. They observed no influence of frequency on the fatigue strength, and no fatigue limit until 1E9 cycles was observed. Similarly, no influence of frequency was observed on aluminum silicon alloy *DISPAL®* in the investigations of Stanzl-Tschegg et al. [176]. These tests were carried out at 140 Hz and 20 kHz at two temperatures – room temperature and 150 °C, and no difference could be identified at any of these temperatures. However, in the study of Zhu et al. [177] on cast aluminum alloy E319, higher fatigue lives at ultrasonic frequency were obtained as compared to 75 Hz. In totality, fcc alloys show less sensitivity to the influence of strain rate; however, the investigations need to be further carried out for specific material and testing as well as environmental conditions.

In the current case of investigated configurations of AlSi12 alloy, the trend observed in literature for fcc structures is identified. In Fig. 5.19, trend lines show no significant influence of frequency. In batch F validation tests at stress amplitudes of 90 MPa and 100 MPa lie on the line of Woehler curve. For batch H, the test at 80 MPa performed at 20 Hz was suspended at 2E6 cycles. At 110 MPa, the fatigue life at 20 Hz lies right to that obtained at 20 kHz, which is in contrast to the literature inferences where the frequency effect would increase the fatigue life at higher frequencies. Therefore, this deviation can be considered as regular scatter in the fatigue life. It was further investigated by analyzing the data using homogeneity of variance test. Three types of formulations were used for homogeneity of variance test – Levene's test with absolute deviations, Levene's test with squared deviations, and Brown-Forsythe test. Results in terms of F-value and p-value for these tests are given in Table 5.3. A significance level of 0.05 was used to evaluate the results. Levene's test with absolute deviations with a p-value of 0.018 suggests that the variance of the two batches is significantly different; however Levene's test with squared deviations as well as Brown-Forsythe results in a p-value higher than 0.05 suggesting that the batch variance is not statistically significant. It can be interpreted that the deviation can be treated as regular fatigue scatter.

Table 5.3:   Variations in three homogeneity of variance tests of batch H

	Levene's (abs. dev)	Levene's (sq. dev.)	Brown Forsythe
F-value	14.752	3.995	1.179
p-value	0.018	0.116	0.338

5.1.5   Crack propagation behavior

Fatigue crack propagation tests were carried out for batches F and H. Results in the form of crack propagation rate (da/dN) as a function of stress intensity factor range (ΔK) are portrayed in Fig. 5.22. The results are discussed in the three regions of the curve i.e. threshold region, Paris region and region of rapid crack growth. In the threshold region, batch F has a threshold value of stress intensity factor range (ΔK_{th}) 3.2 MPa$\sqrt{m}$ and batch H has this value of 3.5 MPa$\sqrt{m}$.

The values represent a distinct difference in the fatigue crack propagation threshold of the two batches, and suggests that base plate-heated parts can endure higher ΔK without letting fatigue crack to propagate. In the Paris region from the Fig. 5.22, the rate of crack growth for the two batches (F and H) is almost similar as can be observed from the graph and by comparing eq. 5.6 and 5.7.

$$\frac{da}{dN} = -5.69(\Delta K)^{2.737} \tag{5.6}$$

$$\frac{da}{dN} = -5.77(\Delta K)^{2.808} \tag{5.7}$$

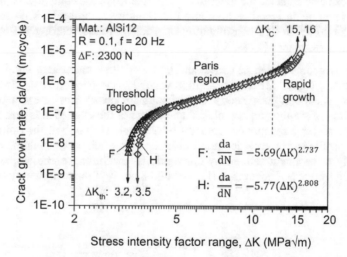

Fig. 5.22: Fatigue crack growth behavior as a function of stress intensity factor range for: without base plate heating batch F, and with base plate heating batch H

In the critical stress intensity factor range region, ΔK_C is 16 MPa√m for the base plate-heated higher than that of batch F which is 15 MPa√m. It is inferred that the resistance to crack propagation can be improved in-process by base plate heating which is attributed to the coarser microstructure (dendritic width 0.56 µm as compared to 0.35 µm). Coarse grain structure is advantageous in crack propagation.

5.2   Hybrid AlSi12 alloy structures [4]

Selective laser melting (SLM) process, having competitive advantage of manufacturing intricate part designs, is especially relevant for manufacturing of parts in machine tool industry – tools, jigs, fixtures, patterns and molds – consisting of complex geometries. The potential of manufacturing such parts by additive manufacturing (AM) is attractive also in the sense that these parts are not only complex but also manufactured in low volumes. Application of additive manufacturing in this industry can therefore reduce lead times which can be a driver for innovative parts, which usually require new toolings and extended product development cycle. Application of additive manufacturing to tooling industry can also enhance the functional performance of the components. As these structures can be designed without any functionality compromise with regard to manufacturability, design consolidation by additive manufacturing will enhance functional performance [178–180].

Though any geometry can be manufactured by additive manufacturing, it should be carefully weighed for its justification related to production costs. Several of these component designs are consisting of not a completely intricate structure, but combination of simple and complex shapes. One of the examples is presented in Fig. 5.23a where a component consists of a simple structure at the bottom and complex profile and internal features at the top part. Such parts would be difficult and costly to be manufactured by conventional machining processes, but their processing by additive manufacturing would be costly if the complete structure is

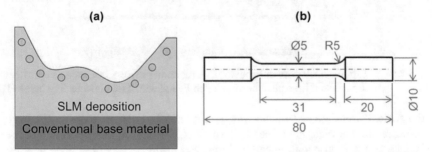

Fig. 5.23:   Schematic of an exemplary hybrid structure (a); machined tensile specimen, dimensions in mm (b) [159]

[4] Results presented in this section are partly published in [149,159,178].

manufactured solely by AM. Here it is possible to combine both types of manufacturing processes in a hybrid process where the lower simple part can be taken from conventionally-manufactured material, and the intricate profile can be generated above that by additive manufacturing.

For the functional applications of these structures in industrial applications, it is important that their mechanical performance be ensured. Therefore, this chapter investigates the mechanical performance of AlSi12 alloy additively generated by SLM process on two different conventionally-manufactured base materials (Al7020 and Al6082). Manufacturing was carried out at different processing conditions. Their joint strength has been compared by tensile tests, by carrying out different post-processing to achieve the optimal strength and toughness values. Best performing alloy under quasistatic loading has been tested in fatigue until VHCF range, as such components undergo cyclic loading until very high number of cycles.

5.2.1   Specimen manufacturing and post-processing

Conventionally-manufactured alloys Al7020 and Al6082 were taken as base materials and AlSi12 alloy was additively melted above them according to the concept elaborated in Fig. 5.23a. Specimens were manufactured according to the processing conditions listed in Table 5.4 and machined afterwards to the geometry for tensile testing in Fig. 5.23b, for HCF as in Fig. 3.6 and for VHCF as given in Fig. 3.12 such that the transition between the base material and additively-manufactured part remains in the middle of the specimen. Stress-relief at 240 °C for 2 hrs was carried out to remove the induced residual stresses during processing. Following design of experiments (Table 5.5) was designed to investigate the influence of different post-process conditions on the hybrid specimens.

Table 5.4:   Processing conditions for manufacturing hybrid specimens

Parameter set	Scan speed [mm/s]	Laser power [W]	Energy density [J/mm³]
1	930	350	44.28
2	930	350	44.28 (double exposure to 1st layer)
3	620	350	66.41 (for first 10 layers) / 44.28

Table 5.5:   Design of experiments for quasistatic tensile investigations

Batch	Process *	Material	Heat treatment
M	Conv.	Al7020	T6
N	Conv.	Al7020	T6 + SR †
P	Conv.	Al6082	T6
Q	Conv.	Al6082	T6 + SR
R	Hybr.	Al7020 + AlSi12	-
S	Hybr.	Al7020 + AlSi12	SR
T	Hybr.	Al6082 + AlSi12	-
U	Hybr.	Al6082 + AlSi12	SR

* Conv. = conventionally-manufactured; Hybr. = Hybrid
† SR = stress-relief

5.2.2   Interface characterization

Exemplary surface micrographs at the transition region of Al7020 alloy with parameter set 2 and 3 are shown in Fig. 5.24a and b respectively. Interface between the two materials show good joining of AlSi12 alloy melted by SLM process and Al7020 conventional alloy. Fusion of the alloys is recognizable in a small region between 100 and 150 μm which corresponds to thickness of two to three layers in the SLM process. The micrographs do not show any signs of difference in melting for the two parameter sets.

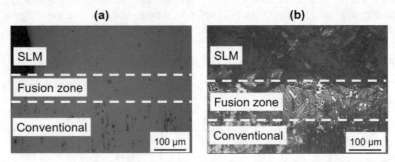

Fig. 5.24:   Transition region in melting of AlSi12 above conventional Al7020 with parameter set 2 (a); and parameter set 3 (b) [178]

The two specimens were investigated for hardness in the transition zone before and after stress- relief at 240 °C for 2 hrs and the results are plotted in Fig. 5.25. Vertical dashed line in the figure is the transition plane, the left of which is the conventional Al6082 alloy, and the right side of the plane is AlSi12 alloy additively-generated by SLM process. In a few initial SLM layers, there exists a small scatter in the hardness values, which is attributed to the interface layers and is stabilized afterwards.

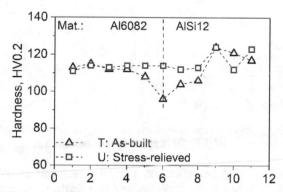

Fig. 5.25:   Hardness profile around the fusion zone for melting of AlSi12 above conventional Al6082

5.2.3   Quasistatic behavior

Tensile tests were carried out for the configurations listed in Tbel 5.5. Fig. 5.26a portrays the influence of hybridization with AlSi12 alloy on the quasistatic tensile properties for Al7020 and Al6082. Al7020 under T6 condition (Batch M) has a tensile strength of 345 MPa and fracture strain of more than $14.0 \cdot 10^{-2}$. Al6082 under the same condition (Batch P) has tensile strength of 420 MPa and fracture strain of $3.5 \cdot 10^{-2}$. After hybridizing these alloys with AlSi12 alloy with SLM process, the fracture strain of Al7020 decreases from $14.0 \cdot 10^{-2}$ to $4.5 \cdot 10^{-2}$ (Batch R) and that of Al6082 from $3.5 \cdot 10^{-2}$ to $2.0 \cdot 10^{-2}$ (Batch T). The change in tensile strength after hybridization is dissimilar for the two investigated alloys with the major effect remains as a shift towards brittle behavior with significant decrease of fracture strain for both the alloys.

All the specimens were tested also after stress-relieving heat treatment and the results are plotted in Fig. 5.26b. For Al7020 pure alloy, ultimate tensile strength reduced from 345 MPa to 330 MPa with a slight increase in fracture strain from $14.0 \cdot 10^{-2}$ to $14.5 \cdot 10^{-2}$ (Batch M and Batch N respectively). For Al6082 alloy, stress-relief decreased the ultimate tensile strength from 420 MPa to 340 MPa; whereas fracture strain increased from $3.5 \cdot 10^{-2}$ to about $7.0 \cdot 10^{-2}$ (batch P and Q respectively).

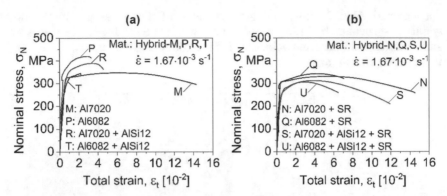

Fig. 5.26:   Characteristic stress-strain curves showing the effect of: hybridization (a); and stress-relief (SR) for pure and hybrid batches (b)

Stress-relief had a significant effect on the fracture strain of hybrid batches. Al7020 + AlSi12 specimens (Batch S) reached to a fracture strain of $11.9 \cdot 10^{-2}$ as compared to $4.5 \cdot 10^{-2}$ without stress-relief. Al6082 + AlSi12 alloy increased its fracture strain from $2.0 \cdot 10^{-2}$ to $6.5 \cdot 10^{-2}$ after stress-relief. The achieved values of fracture strain are only slightly less than those of pure alloys. The relatively small decrease in ultimate tensile strength after stress-relief is accompanied by a large increase in fracture strain of the hybrid specimens.

Significantly improved ductility of the hybrid specimens after stress-relief is due to the fracture behavior: before stress-relief, all the hybrid specimens broke from the transition zone; whereas after stress-relief, the region of fracture was shifted to the pure alloy i.e. Al7020 and Al6082 which increased the fracture strain. Fracture from the joining plane is an indicator of the presence of residual stresses at the fusion zone, which are expected to be removed after stress-relief such that the joint strength is more than the strength of the pure alloy.

Fracture surfaces, observed under scanning electron microscope, show the behavior of tensile fracture. Specimens without stress-relief failing from the joining plane is characterized by brittle fracture from the defects (Fig. 5.27a) resulting in sufficiently reduced fracture strain. Fracture surface of hybrid specimen with Al7020 in Fig. 5.27b is characteristic of a ductile fracture with sufficient plastic deformation seen in the ridges of the fractograph resulting in improved fracture strain. Intermediate fracture strain in Al6082 specimens is characterized by the relatively brittle fracture (Fig. 5.27c) due to presence of a small amount of Si in the material.

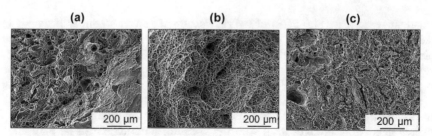

Fig. 5.27: Fracture surfaces of tensile hybrid specimens with Al7020 batch R (a); Al7020 with stress-relief batch S (b) [159]; Al6082 with stress-relief batch U (c)

5.2.4 Fatigue behavior in HCF range

Based on the quasistatic properties, batch S (Al7020 + AlSi12 with stress-relief) was selected to be tested under fatigue loading. Analysis of fatigue strength of parts was carried out according to sections 0 and 3.6 for HCF and VHCF respectively. Fig. 5.28 portrays the results of load increase test for the hybrid specimen (batch S). Results of AlSi12 pure alloy conditions (batches E,F,H) are also displayed together for reference.

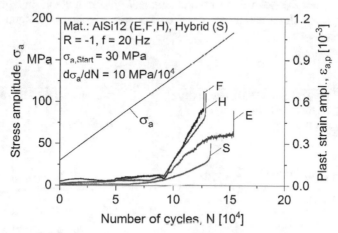

Fig. 5.28: Exemplary course of material response in load increase test of hybrid specimen Al7020 + AlSi12 followed by stress-relief batch S, together with pure AlSi12 configurations E, F and H: as-built batch E, with stress-relief batch F, with base plate heating and stress-relief batch H

Load increase test (Fig. 5.28) of hybrid specimen shows a similar behavior to AlSi12 batches in the initial stress loading and the critical stress range starts after about 110 MPa. The induced plastic strain amplitude until fracture $(0.2 \cdot 10^{-3})$ experienced in hybrid specimen remains less as compared to that in pure alloy conditions $(0.4 \cdot 10^{-3}\text{-}0.6 \cdot 10^{-3})$. The effect of decreased plastic strain has to be confirmed by testing at constant stress amplitude. From the load increase test, stress amplitude of 120 MPa was selected for constant amplitude tests which also makes it possible to compare it with the constant amplitude tests of pure alloy conditions.

Cyclic deformation behavior at 120 MPa of an exemplary hybrid specimen is shown in Fig. 5.29a. For reference, corresponding curves for the SLM-manufactured batches, discussed in section 5.1.3 are also plotted. Following the trend observed in load increase test (Fig. 5.28), hybrid specimen exhibits less magnitude of plastic strain amplitude as compared to the pure alloy conditions. At high stress amplitude of 140 MPa, shown in Fig. 5.29b, the specimen shows similar behavior with increased value of plastic strain amplitude. At both the stress amplitudes, specimens experience continued cyclic hardening until fracture. Cyclic hardening in hybrid specimens comes partly from the SLM process, as experienced in batch F. Additionally, it can be attributed to fusion of the two alloys at the interface between conventional and additive manufacturing plane. The reduced plastic deformation has resulted consistently higher fatigue lives for hybrid specimens.

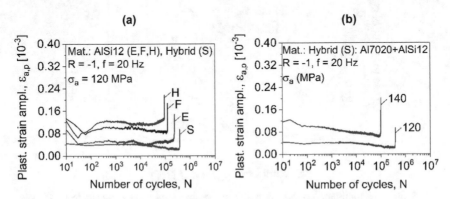

Fig. 5.29:   Cyclic deformation behavior of hybrid specimen batch S at: 120 MPa together with AlSi12 batches E, F and H for reference (a); 120 MPa and 140 MPa (b)

Results of constant amplitude tests at 120 MPa are plotted in Fig. 5.30, where hybrid specimens (batch S) show increased fatigue strength together with an increased fatigue scatter. Hybrid batch is to be compared with batch F, as both were manufactured under same conditions. The specimens broke from SLM side of the material as seen in Fig. 5.31. The difference in fatigue life of different specimens corresponds to the defect distribution inside the respective specimens.

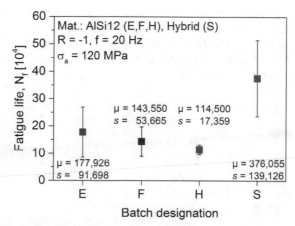

Fig. 5.30: Average fatigue life (μ) and fatigue scatter (s) obtained by constant amplitude fatigue tests at 120 MPa of hybrid specimens Al7020 + AlSi12 followed by stress-relief batch S, together with pure AlSi12 configurations E, F and H: as-built batch E, with stress-relief batch F, with base plate heating and stress-relief batch H

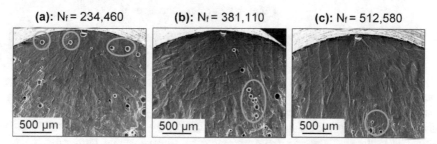

(a): N_f = 234,460   **(b):** N_f = 381,110   **(c):** N_f = 512,580

Fig. 5.31: Fractographs explaining the difference in fatigue life depending on the location of pores at 120 MPa of the hybrid specimens batch S

Fig. 5.31a shows the specimen which endured the least fatigue life which corresponds to the existence of several material defects in the vicinity of the specimen surface. Fig. 5.31b represents specimen having limited defects near the surface and the specimen endured intermediate life. Specimen having defects far from the surface were not susceptible to early fracture from these defects, as represented in the fractograph in Fig. 5.31c.

5.2.5   Fatigue behavior in VHCF range

Very high cycle fatigue tests were also carried out for hybrid specimens i.e. batch S as the potential applications of such structures can range to very high number of load cycles. Fig. 5.32 represents the combined results of fatigue testing of hybrid specimens carried out at servohydraulic testing system (f = 20 Hz) and ultrasonic fatigue testing system (f = 20 kHz).

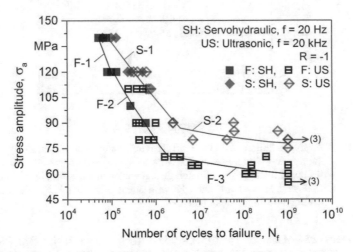

Fig. 5.32:   Woehler curves for hybrid alloy Al7020 + AlSi12 batch S, together with pure AlSi12 alloy batch F until very high cycle fatigue range

The results of hybrid batch are compared with the results of the corresponding pure alloy condition i.e. batch F. Results show that hybrid specimens consistently outperform the pure alloy throughout the stress range. This higher strength can be referred to the decreased plastic deformation in the load increase test which suggests that the specimens have experienced relative cyclic hardening. Better joining in the transition zone by stress-relief can also be reason of higher performance which has shifted the mechanism of fatigue failure to outside the smallest cross sectional area. All the specimens failed from the SLM side, and fractography showed a similar role of process-induced defects as in the pure alloy.

Regarding the fitting of the Woehler lines in different ranges, batch F was already discussed in section 5.1.4. The fitting equations for fatigue life (N_f) in terms of applied stress amplitude for batch F are given for its three ranges in eq. 5.8 to 5.10.

Though hybrid specimens were not manufactured under heating of base plate, they also show only one transition point like batch H i.e. transition from HCF to VHCF ranges. The fitting equations for the hybrid batch for the two identified ranges are given as eq. 5.11 and 5.12. The slope factor for hybrid batch is 2.72 (S1:S2) which is in-between the values for batch F (2.82) and batch H (2.67).

$$N_f = (\frac{\sigma_a}{1296})^{-\frac{1}{0.204}} \qquad \text{range F-1} \tag{5.8}$$

$$N_f = (\frac{\sigma_a}{794})^{-\frac{1}{0.164}} \qquad \text{range F-2} \tag{5.9}$$

$$N_f = (\frac{\sigma_a}{188})^{-\frac{1}{0.058}} \qquad \text{range F-3} \tag{5.10}$$

$$N_f = (\frac{\sigma_a}{427})^{-\frac{1}{0.098}} \qquad \text{range S-1} \tag{5.11}$$

$$N_f = (\frac{\sigma_a}{164})^{-\frac{1}{0.036}} \qquad \text{range S-2} \tag{5.12}$$

The trends for the three batches can also be related to the fatigue strength at 1E9 cycles which for hybrid batch is in-between the values of batch F and batch S, but in the opposite order. The susceptibility of batch F for steeper curves can be related to more dependence of batch F on stress level, and has higher rate of decay which can be related to more susceptibility of process-induced pores with very less possibility of plastic deformation. In the base plate-heated batch (H), the pores are reduced, as well as their effect is normalized due to relatively higher grain sizes. Hybrid specimens show the intermediate behavior due to the intermediate structure of the joining alloys.

Frequency effect

Frequency effect for the batch F is already discussed in section 5.1.4, where it was concluded that no frequency effects were identified. For batch S, the instances at stress amplitudes of 110 and 120 MPa show a uniform slope of the Woehler curve. Further statistical analysis was realized for stress amplitude of 120 MPa. Similar analysis as in section 5.1.4 was performed for hybrid batch at 120 MPa. Table 5.6 shows the analysis results for Levene's test with absolute deviations, Levene's test with squared deviations, and Brown-Forsythe test performed to determine the homogeneity of variance at frequencies of 20 Hz and 20 kHz.

Table 5.6:   Variations in three homogeneity of variance tests for batch S

	Levene's (abs. dev)	Levene's (sq. dev.)	Brown Forsythe
F-value	0.009	0.022	0.005
p-value	0.928	0.889	0.944

In the case of hybrid specimens, all the three tests result in p-value higher than 0.05, which recommends that the variance in the fatigue tests performed at 20 Hz and 20 kHz does not exist. Correspondingly, analysis of variance (ANOVA) was performed which also suggested, with p-value 0.278 (F-value: 1.569), that the mean fatigue life of the two groups of tests performed at different frequencies are statistically not different. Therefore, the effect of frequency for these material conditions can be ignored.

5.3   Fatigue prediction methodology [5]

5.3.1   Fracture mechanics-based approach

The pores resulting from the SLM process have been observed to have critical influence on the mechanical behavior of SLM parts. Influence of size of the pore as well as its location on the damage mechanism was investigated for development of plastic strain as the response parameter to applied stress. Pores with different size and distance from surface were analyzed and the equivalent plastic strain (PEEQ) is shown in Fig. 5.33 where PEEQ value is represented for two selected pores – for a pore size of 110 µm located at a distance of 490 µm from outer boundary of the specimen; and for a pore of size 90 µm at a distance of 165 µm from specimen boundary. In 90 µm pore, PEEQ value is higher than that in the pore of 110 µm. The results show that the pores near to surface, even when smaller in size, are critical for damage development.

Previous results in fatigue testing (sections 5.1 and 5.2) have shown that fatigue scatter is caused by multiple crack initiation as well as crack initiation from internal pores. Therefore, it is important to identify the critical pores, which was carried out by calculating stress concentration factor (K_t) for the pores existing in the tomographic model which was meshed afterwards. Fig. 5.34a portrays the stress concentration factor for the pores with a nominal dimeter of 90 µm, which shows that the pores near to surface are more sensitive to applied loading. These pores with high stress concentration factors will be favorable sites for crack initiation and will result in early fracture. Similar-sized pore at a larger distance inside the core will have less chances of causing crack initiation. Therefore, pore diameter (d) was normalized by the relevant distance from surface (ρ), and stress concentration factor was calculated against a newly introduced factor pore characteristic (γ) as a ratio of pore diameter to distance from surface (d/ρ). This graph is shown in Fig. 5.34b which was modeled mathematically.

The model shows exponential growth of K_t with increasing value of pore characteristic. The increase in K_t remains only marginal after a specific value of pore characteristic (0.5). Randomness of residuals and the value of coefficient of determination ($R^2 = 0.997$) explains the convergence of the model; the resulting mathematical model is given in eq. 5.13. Another specimen was used for validation of this model. The difference in the K_t calculated from finite element analysis and

[5] Results presented in this section are partly published in [129,149,155,181].

those predicted from this model were calculated to find out the model accuracy to be in the range of 3%.

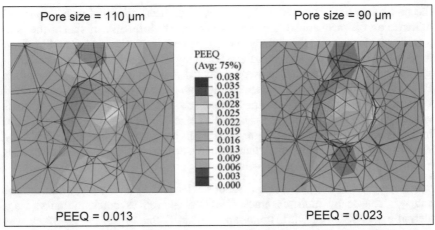

Fig. 5.33: Development of equivalent plastic strain for batch H at two load levels: for largest pore in the specimen (a), and for critical pore near the specimen surface (b) [129]

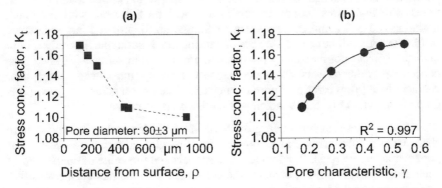

Fig. 5.34: Stress concentration factor (K_t) as a function of: distance from specimen surface (a); and pore characteristic i.e. ratio of pore diameter to its distance from specimen surface (b) [129]

$$K_t = 1.178 - 0.066 \cdot e^{-(\gamma - 0.182)/0.156}$$ (5.13)

Therefore this formulation was used for the defects determined by μ-CT and methodology explained in section 3.8, and the analysis was carried out to find out the most probable fatigue life. The predicted results of batches F and H are plotted together with experimental results in Fig. 5.35. When predicted values are compared against experimental data in the Fig. 5.35, deviations are realized. Deviations tended to appear more vigorously as applied stress level went down.

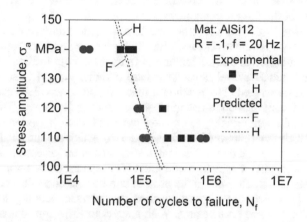

Fig. 5.35: Experimental and stochastically predicted fatigue lives for batch F without base plate heating, and batch H with base plate heating [155]

This phenomenon is believed to be inherent from Weibull's weakest-link theory assumption, which states that structural defects are assumed to be randomly distributed and widely spaced enough, not to allow any interference to occur. It seemed like such assumptions does not hold in this analysis, also investigated by Leuders et al. [182] for Ti-6Al-4V. Being aware of the fact that in low stress high cycle fatigue range, crack initiation phase is the dominant phase in fatigue life of a component, interference of defects not widely spaced enough may lead to deviation from predicted values in favor of higher fatigue life of a component.

Accounting to those phenomena and facts, a justification is proposed from an assumed pore interference scenario named as pore-blunted crack. When two pores occur to be in the vicinity of each other; assuming pore I is to be more critical with higher pore characteristic γ, then a crack is assumed to propagate form pore I in the direction of pore II, after its crack initiation phase is complete. When the crack initiated from pore I reaches pore II, the crack is blunted and a process of crack initiation has to start over, further prolonging the fatigue life of the specimen. Such discrepancies should be taken into account before application of a prediction methodology for industrial components, and should be integrated to the plasticity parameters for the materials which is the topic of plasticity-based approach below.

5.3.2   Plasticity-based approach

The calculation approach, presented in the flow diagram in section 3.8, combining plasticity, finite element method and Monte-Carlo modeling is applied to SLM-manufactured materials. FEM provides a discrete probability density function for the outcomes of the fatigue phenomena per respective stress amplitude. From this density function, instances are generated to anticipate possible outcomes of the Monte-Carlo process. Nevertheless, FEM data on its own gives highlights of the damage accumulation and stress buildup of fatigue loading. In Fig. 5.36, a z-cross section of a batch H model is shown where a radially varying sectional stress distribution is envisaged. The maximum macroscopic stress field can be seen in sub-layers to the surface up to one third of the depth to the center. This yields the zone with its inherent porosity, as the most potential crack initiation domain. This consequently gives rise to multi crack initiation in which the most trivial scenario is one to the surface and one to the center. Both are in the direction of high stress intensity, as a result of which, a compliance of specimen is observed towards imminent fracture. Such event in the process can be related to the secondary sub-surface crack, which has already reached the surface, leading to significant increase of the cracked surface area. With respect to FEM, prelude towards such phenomenon can be observed in Fig. 5.36 where a stress concentration factor amounting to 2.17 is experienced. In that case, a discrete difference between the nominal applied stress to the specimen cross-section, and the stress distribution within the specimen can be clearly realized, which shows that micro-scale porosity, experimentally characterized to be in the order of up to hundreds of microns, can develop micro-stresses in their vicinity amounting to more than double the applied nominal stress amplitude. In Monte-Carlo modeling, the aim is to build a relation between the stress distribution under cyclic loading and the fatigue strength via a deterministic equation modified to accommodate probability variations.

In Fig. 5.37, a cross-section of batch F is envisaged, but since pores were almost evenly distributed on the cross-section, no such sectional-inhomogeneous stress distribution can be seen. This brings to attention the effect of mutual location of pores with respect to each other and the specimen surface and core. Nonetheless, the developed stress concentration due to a nominal amplitude of 80 MPa did amount to 2.28, which is higher than 2.17 of batch H at 110 MPa. This means the relation between concentration factors and stress amplitude is not monotonically linear; it will also experience variation, subject to specific structure. In this regard, strain rate accompanying effects should also influence development of stress concentration.

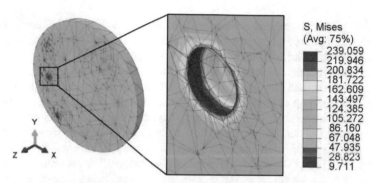

Fig. 5.36: Stress distribution at mid cross-section of a pore in batch H built with base plate heating at 110 MPa of stress amplitude

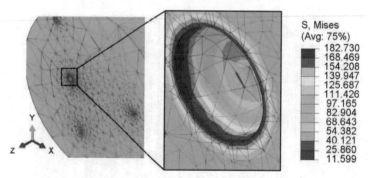

Fig. 5.37: Stress distribution near mid-cross section of a pore in batch F built without base plate heating at 80 MPa of stress amplitude

In the case of hybrid specimen, the stress concentration factor reached up to 2.24, which lies in-between that of batch F (2.28) and batch H (2.17). This comes in context of different cyclic hardening characteristics as a result of the hybrid nature of the structure and the superior plastic properties of the base alloy Al7020. The higher stress concentration factors do not necessarily mean lower fatigue life; it only signifies high potential crack initiation sites, while cycles consumed to initiation of crack is influenced by micro-mechanical plastic properties; however, homogeneities causing stress concentration alter the micro-mechanical response to applied stress, which is significantly influenced by the microstructural features.

Conducting cyclic loading simulations in FEM has shown the extent of field variation that can be introduced to a Monte-Carlo model. In the stochastic simulation, variation of stress did result from geometrical and hardening non-linearities, which was related to fatigue life by an iterative scheme of random event generation and probability assessment. This strategy makes it possible not only to compute a certain fatigue life, but also to find out the fatigue strength variation with respect to a certain applied stress amplitude. The connection between Monte-Carlo simulation and the fatigue life is achieved through application of the probabilistic Basquin equation of eq. 3.19. Additionally, the Monte-Carlo process is designed to account for the random variation expected in the material strength. Damage developed in Markov chain for the respective material configuration is used in parallel to calculate correction factors of material residual strength in eq. 3.20 to eq. 3.22 respectively for different fatigue ranges. The correction also takes into account the developed Dang-Van stress amplitude. Consequently, the computation procedure has accounted for expected deviation of fatigue strength as of strength variation or within specimen stress distribution.

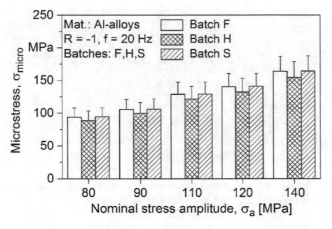

Fig. 5.38:   Comparison between micro-stress (σ_{micro}) values developed inside fatigue specimens at selected stress amplitudes for: AlSi12 with stress-relief batch F, AlSi12 with base plate heating and stress-relief batch H, and Al7020 + AlSi12 with stress-relief batch S

Development of micro-stresses and equivalent plastic strains in batches F, H and S is presented in Fig. 5.38 and Fig. 5.39 respectively, where it is observed that batch H developed less average value of micro-stresses along the applied load spectra of stress amplitudes. Meanwhile, batch F and batch S have almost equivalent micro-stress values.

In fatigue damage, application of load is translated into material's microstructure as plastic damage, therefore the strain developed at an applied stress level can be used as an assessment of degree of damage within a certain structure corresponding to the distribution of plastic strain.

In Fig. 5.39, distribution of micro-plastic strain in terms of equivalent plastic strain (PEEQ) at the applied stress amplitudes for the three batches can be envisaged. For batch F, the behavior observed with increasing stress amplitude is non-monotonic, which is precipitated from non-monotonic cyclic hardening behavior. For batch H, the development of equivalent plastic strain is corresponding to the increase of applied stress amplitude. The gradient of equivalent plastic strain can be considered as a measure of the gradient in fatigue life. For instance, it can be foreseen that the difference in fatigue life of batch H at 120 and 140 MPa of stress amplitudes is smaller as compared to that between 110 and 120 MPa. Difference in the gradients is observed between 90 MPa and 110 MPa, which can be related to the change of slope i.e. knee-point in the Woehler curve.

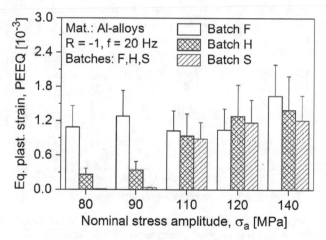

Fig. 5.39:   Comparison between equivalent plastic (PEEQ) values developed inside fatigue specimens at selected stress amplitudes for: AlSi12 with stress-relief batch F, AlSi12 with base plate heating and stress-relief batch H, and Al7020 + AlSi12 with stress-relief batch S

As for difference between 80 and 90 MPa, it reflects a saturation of fatigue life at this level of load in the VHCF range i.e. a small slope of the Woehler curve. Similar behavior is observed for batch S, where knee-point is expected between 90 and 110 MPa. Below 110 MPa, a very large increase in the fatigue life can be foreseen due to a sharp drop of plastic strain. At this level, the fatigue life is well into the VHCF range since corresponding developed plastic strain is very less.

Considering all three batches, batch F developed highest plastic strain deviations, therefore higher fatigue scatter in comparison to batch H and batch S. Such attributes can be justified by resulting strain localization form high strain deviation. Fatigue damage mechanism is firstly a process of strain saturation, consequently localization. In a stochastic sense, it reflects amount of deviation from mean values.

The distribution of stresses resulting from FEM simulations were submitted to Monte-Carlo modeling. Discrete values were employed as discrete probability density functions, from which random samples were drawn and event probabilities assessed, the result of which was fatigue life prediction with lower and upper bounds. The lower bound of fatigue life value was taken corresponding to maximum event probability as shown exemplarily in Fig. 5.40.

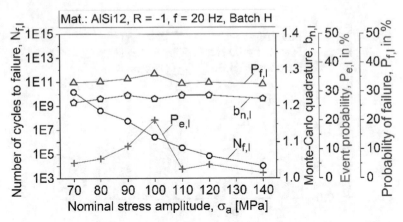

Fig. 5.40:   Projection of statistical response on predicted life at the lower bound of Monte-Carlo simulation (exemplary values for batch H)

It can be seen that Monte-Carlo quadrature value used for the time integration process cannot be considered constant and would vary with respect to the applied stress amplitude in the calculation. Event probability, which is also varying in value, indicates confidence in the prediction upon which design safety factors can be considered. At the point in the Monte-Carlo simulation where first possible fatigue failure is detected, the Markov process is proceeded until the first impossible event in the Monte-Carlo process is detected, which accounts for the upper bound in fatigue life.

Combined lower bounds and upper bounds of predicted fatigue lives for the batches F, H and S are plotted in Fig. 5.41. For batch H, the S-N relation is a smooth decay of fatigue strength. For batch S, the slope of the response parameter from load increase test at 70 MPa did not make it possible to predict probabilities and fatigue life, therefore, prediction was started at 80 MPa of stress amplitude.

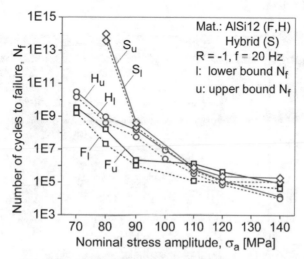

Fig. 5.41:   Predicted lower and upper bounds of fatigue life for: AlSi12 with stress-relief batch F, AlSi12 with base plate heating and stress-relief batch H, and Al7020 + AlSi12 with stress-relief batch S

At a stress amplitude of 110 MPa, the lowest event probability was recorded which came along with highest Monte-Carlo quadrature. The maximum life was in excess of 1E13 cycles at 80 MPa as a lower bound with highest event probability. At the upper bound, highest Monte-Carlo quadrature was at 110 MPa of stress amplitude, meanwhile highest event probability shifted to 140 MPa of stress amplitude. In the same context, maximum fatigue life for the calculated stress amplitudes was beyond 1E14 cycles at 80 MPa. Construction of relationship between analysis outcomes of Monte-Carlo simulation and corresponding statistical responses is important. In that sense, the accuracy of prediction can be judged and safety factors can be set. This is especially important in applications where fatigue reliability is a concern and safety uncertainties have to be accounted for.

A comparison between fatigue life values predicted through the developed prediction methodology and experimental values for batch F is plotted in Fig. 5.42. A good agreement between experimental and computational results can be found between 110 and 140 MPa, all experimental values lie between the upper and lower boundaries of Monte-Carlo model; however, at lower stress amplitudes, experimental values lie outside the prediction interval; at 100 MPa, experimental fatigue life lies at the boundary of the lower limit of predicted fatigue life, and at lower stress amplitudes, the experimental values are lower than the lower limit of the predicted values.

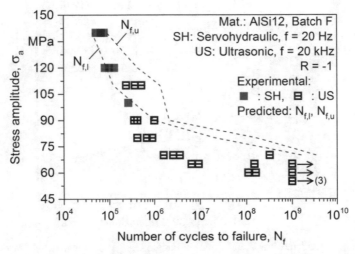

Fig. 5.42:   Comparison between predicted and experimental fatigue lives for AlSi12 with stress-relief batch F

One of the explanations about it can be envisaged is the influence of frequency, experiments at lower stress amplitudes was done at 20 kHz of frequency. This does not agree with the test frequency on which the simulation was based. Load increase tests were held at 20 Hz of frequency, which means 1,000 times difference in strain rate. This comparison highlighted the strain rate sensitivity issue of batch F. Batch F, with finer microstructure than batch H, would provide obstacle to dislocations with more grain boundaries to cross. At sufficiently higher strain rates, grain boundaries suffer substantial reduction of strength and such boundary crossing is possible at lower remote loads. This undermines one of the main microstructural strength mechanisms of the structure of batch F. With load reversal, the mechanism is repeated in compression as well, which gives rise to extensive damage of the structure and subsequently, lower fatigue strength.

Fig. 5.43 shows a comparison between experimental and predicted fatigue lives of batch H. An agreement between predicted and calculated values can be realized along the whole range of applied stress with a limited deviation at 110 MPa, where experimental fatigue life is slightly higher. Although, validation experiments below 110 MPa were also held at 20 kHz of frequency, this did not induce error in the calculation, which shows that the microstructure of batch H is less sensitive to strain rate. Batch H experienced coarser, more ductile structure relative to batch F. Coarser structure with more developed cellular dendrites, is more ductile and tolerant to plastic damage. When cyclic loading is applied, less extensive damage to the microstructure is developed in comparison to batch F. It is only, in low cycle fatigue, higher than 110 MPa, that batch F offers better performance, since finer microstructures will increase resistance to crack initiation as it is harder to develop a slip band. This means that a substantial increase in HCF as well as VHCF strength can be achieved through base plate heating at the expense of low cycle fatigue strength. This highlights the issues of strain rate sensitivity in SLM structures built at higher thermal gradients (batch F). Due to microstructure heterogeneity and sensitivity to thermal history, significant modifications to properties can be achieved by post-SLM treatments.

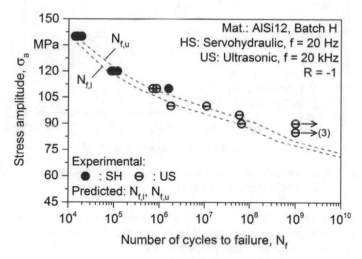

Fig. 5.43:   Comparison between predicted and experimental fatigue lives for AlSi12 with base plate heating and stress-relief batch II

For the hybrid specimens of batch S, experimental fatigue lives and predicted fatigue life interval is presented in Fig. 5.44. Since batch S is of hybrid nature (SLM AlSi12 deposited on Al7020), interface strength between the two materials did influence the total strength of the structure. For an applied stress range between 110 and 140 MPa, an agreement between experimental and predicted fatigue life can be seen; however, below stress amplitude of 110 MPa, experimental fatigue lives lie below the lower limits of predicted fatigue lives with a higher scatter.

The deviation can be attributed to possible variation of joint strength under non-controlled deposition process, which implies that joining defects contributed largely to the fatigue strength variation per applied stress amplitude; even then, fatigue strength of batch S (hybrid) was higher than that of batch F (pure SLM). This highlights potential of manufacturing high strength structures that respect economic constraints.

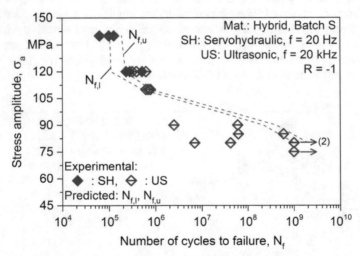

Fig. 5.44:   Comparison between predicted and experimental fatigue lives for Al7020 + AlSi12 with stress-relief batch S

With this comparison, the need for separate attention to joint strength that is developed at the interface was highlighted. Thermal history driving the evolution of the microstructure in this zone would be significantly different. Microstructure is finer which means higher crack initiation resistance, but also the deposition process develops interface defects that can significantly reduce fatigue life and cause premature fracture.

6   Summary

The current research aimed at determination of the mechanical behavior of AlSi12 alloy (Al4047) manufactured by SLM process, so that the potential of manufacturing tailored structures can be utilized to get the competitive advantages of SLM process in the era of fourth industrial revolution. These advantages include light-weighting potential, freedom-of-design, functional enhancement as well as drastically reduced time to market. It is important to understand the influence of different processing and post-processing conditions on the material properties in terms of their physical and microstructural features so that the corresponding mechanical properties can be controlled by controlling the processing conditions. Therefore, this study has explored the complete profile of the processing, post-processing, microstructural and mechanical properties for AlSi12 alloy as well as hybrid structures.

Powder material after the SLM process has a changed shape from spherical for fresh, and mixed spherical and eccentric morphology for the recycled powder. There is a slight increase in average particle size from 23 μm to 26 μm after the SLM process. Si content in the powder material decreased slightly which can be attributed to the precipitation of Si at the periphery of semi-molten powder particles. These minor changes in the powder properties are attributed to the limited heat affected zone in the SLM process, which suggests that powder material can be recycled several times without considerable change in material properties.

SLM process imparts a eutectic microstructure consisting of Al matrix and eutectic dendrites. Due to very high thermal gradients in the process, very fine dendrites grow, the size of which depends on the time available for their growth. Base plate heating reduces the thermal gradients in the process and, therefore, results in coarser dendrites. Post-process stress-relief is effective in grain coarsening only when no pre-heating is performed. It is attributed to the recrystallization potential which is more effective without base plate heating which has smaller grain size and higher number of grain boundaries. Precipitates were found distributed in the material due to very high cooling rates in the SLM process, which potentially increase the strength of SLM-manufactured Al-alloys. These relationships between processing and post-processing parameters and the resulting microstructural parameters can be used to monitor these features locally by controlling the thermal conditions in the process.

Optimized process parameters resulted in a relative density of more than 99.7%. The heating conditions in the process are responsible for the remnant porosity. By

© Springer Fachmedien Wiesbaden GmbH, part of Springer Nature 2019
S. Siddique, *Reliability of Selective Laser Melted AlSi12 Alloy for Quasistatic and Fatigue Applications*, Werkstofftechnische Berichte | Reports of Materials Science and Engineering, https://doi.org/10.1007/978-3-658-23425-6_6

pre-heating the base plate, de-gassing of the process gas can be achieved. Also melt pool has a higher density as compared to the gases released during SLM process. As pre-heating results in reduced cooling rates, gas bubbles get more time to complete the convention process and go through the melt pool. Stress-relief slightly increased the porosity due to interaction at high temperatures for the extended period of time. Non-destructive defect detection was also carried out by using μ-CT and the trends similar to the above were obtained. Slight differences in 2D and 3D defect detection was observed, as 2D metallography can only give information about the cut section as compared to complete volumetric information in computed tomography. Hardness of the specimens also varied according to the conditions with base plate heating and stress-relief both decreasing the hardness values. Similar effect of both the parameters were observed for residual stresses where tensile residual stresses, obtained for as-built specimens, were almost eliminated by application of base plate heating and stress-relief. No pattern of hardness change as a function of height from base plate could be observed for this alloy.

Quasistatic tensile properties of optimized specimens are higher as compared to cast alloys in terms of yield strength and ultimate tensile strength. These values are about two to four times higher than those of sand- and die-cast alloy respectively. The higher values correspond to very fine microstructure and precipitation of Si particles. Fine microstructure is also responsible for the decreased fracture strain of SLM-manufactured parts. Stress-relief process slightly decreases the strength values. By base plate heating, strength is decreased and fracture strain is increased which is justified by the microstructural modification in terms of coarsening of dendritic width. Post-processing by hot isostatic pressing eliminates the remnant porosity, recrystallizing the microstructure resulting in decreased strength and sufficiently increased fracture strain. Alloy sintering at half of the optimal energy density have resulted in strength values equivalent to those of cast alloy with sufficiently reduced fracture strain due to existing bonding defects. These sub-optimal parts can therefore be used for applications which require static strength, and cost can be saved for such applications.

Fatigue testing for optimization of the process can be successfully expedited by combining load increase test and constant amplitude tests. Under fatigue loading, even the small remnant porosity is critical in terms of reliability. In the presence of small pores, there can be internal crack initiation form these defects which is detrimental to fatigue strength. Base plate heating has successfully been employed to improve the fatigue reliability considerably, as large-sized intermittent pores are eliminated and favor surface crack initiation. Rough specimens have very low fatigue strength as compared to machined specimens which implies that post-processing for surface improvement is required for SLM parts to be used in fatigue applications.

Microstructural features and process-induced defects are critical parameters for fatigue strength. These features may have exclusive nature of effect in different fatigue ranges. At higher stress amplitudes, small defects become active which remain inactive at lower stress amplitudes. Base plate heating reduces the fatigue life at higher stresses but improves it in the HCF and VHCF ranges together with improved fatigue reliability. Fatigue strength of SLM parts for polished specimens is higher as compared to conventionally-manufactured parts. Fatigue strength of SLM-manufactured alloy at 1E9 cycles was found to be higher than that of the conventionally-manufactured alloy at 2E6 cycles. Crack propagation behavior was also improved by base plate heating.

Additive manufacturing can be used together with the conventionally-manufactured parts to manufacture structures which are partially intricate. Such hybrid structures can reduce the costs of manufacturing while maintaining the benefit of rapid production by SLM process. For the as-built specimens, the joining section tears apart by mechanical loading which can be attributed to the presence of residual stresses at the joint. Appropriate post-process stress-relief improved the joint strength such that the hybrid structures did not fail from the joint both under quasistatic and fatigue loading. Therefore, hybrid manufacturing offers a potential to harness the benefits of conventional as well as additive manufacturing.

It is important to quantify the effects of process-induced defects on the fatigue behavior. For this purpose, application of μ-CT, FE analysis and statistical analysis serve as a good tool for predicting the fatigue life of a part based on its real defect status, simulating it for the loading amplitudes and applying the statistical tools. With the existence of fatigue limit losing its authenticity due to availability of VHCF testing systems, fatigue life was predicted instead of fatigue limit using these techniques. A methodology for fatigue life prediction was developed making use of non-destructive defect analysis, finite element analysis as well as statistical techniques. The technique proved to be in agreement for non base plate-heated batches at higher stress levels, while lower stress levels resulted in over-estimation of fatigue life. For the base plate heated specimens, the prediction methodology was found in agreement with the experimental results along the whole range of applied stresses, which is attributed to the compliant microstructure which removed the strain rate sensitivity. Therefore, a time-efficient prediction of fatigue life can be made for materials and structures, significantly reducing the testing efforts.

7 Outlook

Results from the current study are encouraging that the SLM-manufactured AlSi12 alloy exhibits material properties responsible for improved quasistatic strength. After process optimization, remnant porosity has decreased to a very small level which has improved the fatigue reliability. The phenomena perceived in this study have suggested that several topics in this field have a potential to be investigated as comprehensive projects so that the complete capability profile of the process can be utilized for functional applications.

Functional grading of parts by SLM process is the foremost field of research which can help in several aspects of design e.g. light-weighting, functional enhancements and freedom-of-design. The process has the potential to impart localized properties in the components. Functional grading should be explored in two dimensions i.e. with respect to process-induced defects as well as with respect to microstructural modifications. Section 5.3 has shown how the process-induced defects, even with the small remnant porosity, can influence the development of stress concentration factors. To improve fatigue reliability, critical features of a part can be improved by regional re-melting initially at the specimen level. The additional processing at required locations is expected to restrict the crack initiation from critical pores in the vicinity of the free-surface and improve fatigue reliability. This concept can be extended to the component level as well, where geometrical features are responsible to induce higher stress concentrations, which can be taken into account by regional re-melting.

Microstructural features could be controlled in the SLM process by controlling different processing parameters like laser power and scanning speed. Therefore, multiple microstructures can be produced at selected locations in a component. Components can be manufactured with localized properties depending on the susceptibility to fatigue crack. However, the application of such concepts needs to be investigated for different materials and geometrical conditions. Another feature to be investigated related to functional grading is the control of residual stresses in the process which is related to the cooling conditions. If the effect of cooling conditions is completely understood, residual stresses can be controlled at will, keeping the geometrical and loading conditions in consideration.

Relatively high surface roughness of SLM parts is detrimental for fatigue strength. Machined parts have good fatigue strength, but it may not be possible to perform post-process machining on intricate parts. Other surface modification techniques should be investigated e.g. sand-blasting, electro-polishing as well as coating

© Springer Fachmedien Wiesbaden GmbH, part of Springer Nature 2019
S. Siddique, *Reliability of Selective Laser Melted AlSi12 Alloy for Quasistatic and Fatigue Applications*, Werkstofftechnische Berichte | Reports of Materials Science and Engineering, https://doi.org/10.1007/978-3-658-23425-6_7

techniques. Chemical techniques should be given prime consideration, as they can be more suitable for improving surface conditions of the internal features which are more relevant for the process nature.

Additive manufacturing is favorable for weight-reduction not only by topology optimization, but also due to its potential for manufacturing thin-walled structures. Utilizing the technique, cellular structures are made having low relative density, but higher specific strength. Cellular structures have recently been manufactured by additive manufacturing techniques; however, to utilize them in functional applications, their mechanical behavior needs to be investigated for stretch as well as bending dominated cellular structures for their localized damage development.

The higher mechanical strength of SLM-manufactured AlSi12 alloy is not only due to fine microstructure but also due to formation of precipitates. Therefore, additional modification of existing alloys should be investigated e.g. metal matrix composites (MMC). Manufacturing of these composites by SLM process is expected to give the synergic benefits of MMC and SLM process.

Hybrid structures have exhibited their potential by illustration of comparable properties to pure alloys. This concept has a significant potential in several engineering applications like jigs, molds and, most importantly, tool inserts. Manufacturing of hybrid structures for intricate geometries should be investigated in terms of design rules as well as mechanical behavior. Also combination of subtractive and additive manufacturing should be investigated so that the two technologies complement each other for technological and economic benefits

References

[1]   Gibson, I; Rosen, D; Stucker, B: Additive manufacturing technologies: Rapid prototyping to direct digital manufacturing. ISBN: 978-1-493-94455-2 (2015).

[2]   Kruth, J-P; Leu, M; Nakagawa, T: Progress in additive manufacturing and rapid prototyping. CIRP Annals - Manufacturing Technology, 47, 2 (1998) 525–40.

[3]   Campbell, I; Bourell, D; Gibson, I: Additive manufacturing: Rapid prototyping comes of age. Rapid Prototyping Journal, 18, 4 (2012) 255–8.

[4]   Vaezi, M; Seitz, H; Yang, S: A review on 3D micro-additive manufacturing technologies. The International Journal of Advanced Manufacturing Technology, 67, 5-8 (2013) 1721–54.

[5]   Horn, T; Harrysson, O: Overview of current additive manufacturing technologies and selected applications. Science Progress, 95, 3 (2012) 255–82.

[6]   Wendel, B; Rietzel, D; Kuehnlein, F; Feulner, R; Huelder, G; Schmachtenberg, E: Additive processing of polymers. Macromolecular Materials and Engineering, 293, 10 (2008) 799–809.

[7]   Petrovic, V; Vicente, H; Jordá, F; Delgado, G; Ramón, B; Portolés, G: Additive layered manufacturing: Sectors of industrial application shown through case studies. International Journal of Production Research, 49, 4 (2011) 1061–79.

[8]   Frazier, W: Metal Additive Manufacturing: A Review. Journal of Materials Engineering and Performance, 23, 6 (2014) 1917–28.

[9]   Murr, L; Martinez, E; Amato, K; Gaytan, S; Hernandez, J; Ramirez, D; Shindo, P; Medina, F; Wicker, R: Fabrication of metal and alloy components by additive manufacturing: Examples of 3D materials science. Journal of Materials Research and Technology, 1, 1 (2012) 42–54.

[10]   Siddique, S; Wycisk, E; Frieling, G; Emmelmann, C; Walther, F: Microstructural and mechanical properties of selective laser melted Al 4047. Applied Mechanics and Materials, 752-753 (2015) 485–90.

[11]   Kruth, J-P; Mercelis, P; van Vaerenbergh, J; Froyen, L; Rombouts, M: Binding mechanisms in selective laser sintering and selective laser melting. Rapid Prototyping Journal, 11, 1 (2005) 26–36.

[12]   Gornet, T; Davis, K; Starr, T; Mulloy, K: Characterization of selective laser sintering materials to determine process stability. In: Proceedings of the solid freeform fabrication symposium on characterization of selective laser sintering materials to determine process stability, (2002) 546–53.

[13]   Karapatis, P: A sub-process approach of selective laser sintering [Ph.D. Thesis]. Lausanne: EPFL; (2002).

[14]   Kruth, J-P; Wang, X; Laoui, T; Froyen, L: Lasers and materials in selective laser sintering. Assembly Automation, 23, 4 (2003) 357–71.

© Springer Fachmedien Wiesbaden GmbH, part of Springer Nature 2019
S. Siddique, *Reliability of Selective Laser Melted AlSi12 Alloy for Quasistatic and Fatigue Applications*, Werkstofftechnische Berichte | Reports of Materials Science and Engineering, https://doi.org/10.1007/978-3-658-23425-6

[15]   Kruth, J-P; Froyen, L; Rombouts, M; van Vaerenbergh, J; Mercelis, P: New ferro powder for selective laser sintering of dense parts. CIRP Annals - Manufacturing Technology, 52, 1 (2003) 139–42.

[16]   Kruth, J-P; Froyen, L; van Vaerenbergh, J; Mercelis, P; Rombouts, M; Lauwers, B: Selective laser melting of iron-based powder. Journal of Materials Processing Technology, 149, 1-3 (2004) 616–22.

[17]   Rombouts, M; Kruth, J; Froyen, L; Mercelis, P: Fundamentals of selective laser melting of alloyed steel powders. CIRP Annals - Manufacturing Technology, 55, 1 (2006) 187–92.

[18]   Bremen, S; Meiners, W; Diatlov, A: Selective laser melting: A manufacturing technology for the future? Laser Technik Journal, 9, 2 (2012) 33–8.

[19]   Qiu, D; Langrana, N: Void eliminating toolpath for extrusion-based multi-material layered manufacturing. Rapid Prototyping Journal, 8, 1 (2002) 38–45.

[20]   Hoque, M; Chuan, Y; Pashby, I: Extrusion based rapid prototyping technique: An advanced platform for tissue engineering scaffold fabrication. Biopolymers, 97, 2 (2012) 83–93.

[21]   Choi, J; Chang, Y: Characteristics of laser aided direct metal/material deposition process for tool steel. International Journal of Machine Tools and Manufacture, 45, 4-5 (2005) 597–607.

[22]   Dinda, G; Dasgupta, A; Mazumder, J: Laser aided direct metal deposition of Inconel 625 superalloy: Microstructural evolution and thermal stability. Materials Science and Engineering: A, 509, 1-2 (2009) 98–104.

[23]   Campanelli, S; Contuzzi, N: Capabilities and performances of the selective laser melting process, In: Joo, EM (Ed.) New trends in technologies: Devices, computer, communication and industrial systems, ISBN: 978-9-533-07212-8 (2010) 233-52.

[24]   Thomas, D: The development of design rules for selective laser melting: Computer aided product design [Ph.D. Thesis]. Cardiff: University of Wales; (2009).

[25]   Over, C: Generative Fertigung von Bauteilen aus Werkzeugstahl X38CrMoV5-1 und Titan TiAl6V4 mit "Selective Laser Melting" [Ph.D. Thesis]. Aachen: RWTH; (2003).

[26]   Gebhardt, A: Generative Fertigungsverfahren: Rapid prototyping - rapid tooling - rapid manufacturing. ISBN: 978-3-446-22666-1 (2007).

[27]   Munsch, M: Reduzierung von Eigenspannungen und Verzug in der laseradditiven Fertigung [Ph.D. Thesis]. Hamburg: Technical University Hamburg Harburg; (2013).

[28]   Yadroitsev, I; Pavlov, M; Bertrand, P; Smurov, I: Mechanical properties of samples fabricated by selective laser melting, In: 14èmes Assises Européennes du Prototypage & Fabrication Rapide, Paris (2009).

[29]   Rehme, O: Cellular design for laser freeform fabrication [Ph.D. Thesis]. Hamburg: Technical University Hamburg Harburg; (2010).

[30] Elsen, M: Complexity of selective laser melting: A new optimisation approach [Ph.D. Thesis]. Leuven: Catholic University Leuven; (2007).

[31] Rieper, H: Systematic parameter analysis for selective laser melting (SLM) of silver-based materials [Ph.D.Thesis]. Louisville: University of Louisville; (2013).

[32] Glardon, R; Karapatis, N; Romano, V; Levy, G: Influence of Nd:YAG parameters on the selective laser sintering of metallic powders. CIRP Annals - Manufacturing Technology, 50, 1 (2001) 133–6.

[33] Tolochko, N; Khlopkov, Y; Mozzharov, S; Ignatiev, M; Laoui, T; Titov, V: Absorptance of powder materials suitable for laser sintering. Rapid Prototyping Journal, 6, 3 (2000) 155–61.

[34] Herzog, D; Vanessa, S; Wycisk, E; Emmelmann, C: Additive manufacturing of metals. Acta Materialia, 117 (2016) 371–92.

[35] Lewandowski, J; Seifi, M: Metal additive manufacturing: A review of mechanical properties. Annual Reviews of Materials Research, 46 (2016) 151–86.

[36] Yang, Y; Lu, J-B; Luo, Z-Y; Wang, D: Accuracy and density optimization in directly fabricating customized orthodontic production by selective laser melting. Rapid Prototyping Journal, 18, 6 (2012) 482–9.

[37] Ilčík, J; Koutný, D; Paloušek, D: Geometrical accuracy of the metal parts produced by selective laser melting: initial tests, In: Ševčík, L (Ed.) Modern methods of construction design. ISBN: 978-3-319-05202-1 (2014) 573-82.

[38] Muresan, S; Balc, N; Pop, D; Fodorean, I; Radu, S: Manufacture the parts by selective laser melting and their dimensional accuracy. ACTA Technica Napocensis, 55, 1 (2012) 207–10.

[39] Singh, S; Sachdeva, A; Sharma, V: Investigation of dimensional accuracy / mechanical properties of part produced by selective laser sintering. International Journal of Applied Science and Engineering, 10, 1 (2012) 59–68.

[40] Yan, C; Shi, Y; Yang, J; Liu, J: Investigation into the selective laser sintering of styrene–acrylonitrile copolymer and post-processing. The International Journal of Advanced Manufacturing Technology, 51, 9-12 (2010) 973–82.

[41] Fraunhofer IPK: Generative manufacturing methods: Selective laser melting. [April 14, 2017]; Available from: http://www.ipk.fraunhofer.de/en/divisions/production-systems/departments/manufacturing-technologies/selective-laser-melting/.

[42] Kruth, J-P; Badrossamay, M; Yasa, E; Deckers, J: Part and material properties in selective laser melting of metals. Proceedings of the 16th International Symposium on Electromachining (2010).

[43] Rafi, H; Pal, D; Patil, N; Starr, T; Stucker, B: Microstructure and mechanical behavior of 17-4 precipitation hardenable steel processed by selective laser melting. Journal of Materials Engineering and Performance, 23, 12 (2014) 4421–8.

[44] Brandl, E; Heckenberger, U; Holzinger, V; Buchbinder, D: Additive manufactured AlSi10Mg samples using selective laser melting (SLM):

Microstructure, high cycle fatigue, and fracture behavior. Materials & Design, 34 (2012) 159–69.

[45]   Chlebus, E; Kuźnicka, B; Kurzynowski, T; Dybała, B: Microstructure and mechanical behaviour of Ti-6A-7Nb alloy produced by selective laser melting. Materials Characterization, 62, 5 (2011) 488–95.

[46]   Murr, L; Quinones, S; Gaytan, S; Lopez, M; Rodela, A; Martinez, E; Hernandez, D; Martinez, E; Medina, F; Wicker, R: Microstructure and mechanical behavior of Ti-6Al-4V produced by rapid-layer manufacturing, for biomedical applications. Mechanical Behavior of Biomedical Materials, 2, 1 (2009) 20–32.

[47]   Prashanth, K; Scudino, S; Klauss, H; Surreddi, K; Löber, L; Wang, Z; Chaubey, A; Kühn, U; Eckert, J: Microstructure and mechanical properties of Al–12Si produced by selective laser melting: Effect of heat treatment. Materials Science and Engineering: A, 590 (2014) 153–60.

[48]   Vilaro, T; Colin, C; Bartout, J; Nazé, L; Sennour, M: Microstructural and mechanical approaches of the selective laser melting process applied to a nickel-base superalloy. Materials Science and Engineering: A, 534 (2012) 446–51.

[49]   Baufeld, B; Brandl, E; van der Biest, O: Wire based additive layer manufacturing: Comparison of microstructure and mechanical properties of Ti–6Al–4V components fabricated by laser-beam deposition and shaped metal deposition. Journal of Materials Processing Technology, 211, 6 (2011) 1146–58.

[50]   Dinda, G; Dasgupta, A; Mazumder, J: Evolution of microstructure in laser deposited Al–11.28%Si alloy. Surface and Coatings Technology, 206, 8-9 (2012) 2152–60.

[51]   Siddique, S; Imran, M; Wycisk, E; Emmelmann, C; Walther, F: Influence of process-induced microstructure and imperfections on mechanical properties of AlSi12 processed by selective laser melting. Journal of Materials Processing Technology, 221 (2015) 205–13.

[52]   Buchbinder, D; Meiners, W; Wissenbach, K; Poprawe, R: Selective laser melting of aluminum die-cast alloy - Correlations between process parameters, solidification conditions, and resulting mechanical properties. Journal of Laser Applications, 27 (2015) S29205.

[53]   Wauthle, R; Vrancken, B; Beynaerts, B; Jorissen, K; Schrooten, J; Kruth, J-P; van Humbeeck, J: Effects of build orientation and heat treatment on the microstructure and mechanical properties of selective laser melted Ti6Al4V lattice structures. Additive Manufacturing, 5 (2015) 77–84.

[54]   Kempen, K; Thijs, L; van Humbeeck, J; Kruth, J-P: Mechanical properties of AlSi10Mg produced by selective laser melting. Physics Procedia, 39 (2012) 439–46.

[55]   Wycisk, E; Emmelmann, C; Siddique, S; Walther, F: High cycle fatigue (HCF) performance of Ti-6Al-4V alloy processed by selective laser melting. Advanced Materials Research, 816-817 (2013) 134–9.

[56]   Yasa, E; Deckers, J; Kruth, J-P: The investigation of the influence of laser re-melting on density, surface quality and microstructure of selective laser melting parts. Rapid Prototyping Journal, 17, 5 (2011) 312–27.

[57]   Alrbaey, K; Wimpenny, D; Tosi, R; Manning, W; Moroz, A: On optimization of surface roughness of selective laser melted stainless steel parts: A statistical study. Journal of Materials Engineering and Performance, 23, 6 (2014) 2139–48.

[58]   Gebhardt, A; Hötter, J; Ziebura, D: Impact of SLM build parameters on the surface quality. Forum fuer Rapid Technologie (2014).

[59]   Mercelis, P; Kruth, J-P: Residual stresses in selective laser sintering and selective laser melting. Rapid Prototyping Journal, 12, 5 (2006) 254–65.

[60]   Roberts, I: Investigation of residual stresses in the laser melting of metal powders in additive layer manufacturing [Ph.D. Thesis]. Wolverhampton: University of Wolverhampton; (2012).

[61]   Cherry, J; Davies, H; Mehmood, S; Lavery, N; Brown, S; Sienz, J: Investigation into the effect of process parameters on microstructural and physical properties of 316L stainless steel parts by selective laser melting. The International Journal of Advanced Manufacturing Technology, 76, 5-8 (2015) 869–79.

[62]   Stephens, R; Fatemi, A; Stephens, R; Fuchs, H: Metal fatigue in engineering. ISBN: 978-0-471-51059-8 (2000).

[63]   Juijerm, P: Fatigue behavior and residual stress stability of deep-rolled aluminium alloys AA5083 and AA6110 at elevated temperature [Ph.D. Thesis]. Kassel: University of Kassel; (2006).

[64]   Smith, R; Hirschberg, M; Manson, S: Fatigue behavior of metallic materials under strain cycling in low and intermediate life range. National Aeronautics and Space Administartion, Technical Note: D-1574 (1963).

[65]   Landgraf, R; Morrow, J; Endo, T: Determination of the cyclic stress-strain curve. Journal of Materials, 4 (1969) 176–88.

[66]   Basan, R; Franulovic, M; Smokvina Hanza, S: Estimation of cyclic stress-strain curves for low-alloy steel from hardness. Metalurgija, 49, 2 (2010) 83–6.

[67]   Mughrabi, H: Specific features and mechanisms of fatigue in the ultrahigh-cycle regime. International Journal of Fatigue, 28, 11 (2006) 1501–8.

[68]   Spriestersbach, D; Grad, P; Kerscher, E: Crack initiation mechanisms and threshold values of very high cycle fatigue failure of high strength steels. Procedia Engineering, 74 (2014) 84–91.

[69]   Qian, G; Hong, Y; Zhou, C: Investigation of high cycle and very high cycle fatigue behaviors for a structural steel with smooth and notched specimens. Engineering Failure Analysis, 17, 7-8 (2010) 1517–25.

[70]   Pyttel, B; Schwerdt, D; Berger, C: Very high cycle fatigue – Is there a fatigue limit? International Journal of Fatigue, 33, 1 (2011) 49–58.

[71]   Wycisk, E; Solbach, A; Siddique, S; Herzog, D; Walther, F; Emmelmann, C: Effects of defects in laser additive manufactured Ti-6Al-4V on fatigue properties. Physics Procedia, 56 (2014) 371–8.

[72]   Lados, D; Apelian, D: Fatigue crack growth characteristics in cast Al–Si–Mg alloys. Materials Science and Engineering: A, 385, 1-2 (2004) 200–11.

[73]   Turnbull, A; Los Rios, E: The effect of grain size on fatigue crack growth in an aluminium magnesium alloy. Fatigue & Fracture of Engineering Materials and Structures, 18, 11 (1995) 1355–66.

[74]   Schijve, J: Fatigue of structures and materials. ISBN: 978-1-402-06807-2 (2009).

[75]   James, M; Hughes, D; Chen, Z; Lombard, H; Hattingh, D; Asquith, D; Yates, J; Webster, P: Residual stresses and fatigue performance. Engineering Failure Analysis, 14, 2 (2007) 384–95.

[76]   Zhuang, W; Halford, G: Investigation of residual stress relaxation under cyclic load. International Journal of Fatigue, 23 (2001) 31–7.

[77]   Basquin, O: The exponential law on endurance tests. American Society for Testing and Materials Proceedings, 10 (1910) 625–30.

[78]   Gope, P: Determination of minimum number of specimens in S-N testing. Journal of Engineering Materials and Technology, 124, 4 (2002) 421.

[79]   ISO: Metallic materials - Fatigue testing - Statistical planning and analysis of data, 77.040.10, 12107:2012; (2012).

[80]   Walther, F: Microstructure-oriented fatigue assessment of construction materials and joints using short-time load increase procedure. Materials Testing, 56, 7-8 (2014) 519–27.

[81]   Walther, F; Eifler, D: Fatigue life calculation of SAE 1050 and SAE 1065 steel under random loading. International Journal of Fatigue, 29, 9-11 (2007) 1885–92.

[82]   Walther, F; Eifler, D: Local cyclic deformation behavior and microstructure of railway wheel materials. Materials Science and Engineering: A, 387-389 (2004) 481–5.

[83]   Starke, P; Walther, F; Eifler, D: "PHYBAL" a short-time procedure for a reliable fatigue life calculation. Advanced Engineering Materials, 12, 4 (2010) 276–82.

[84]   Polák, J: Persistent slip bands (PSBs), In: Wang, C (Ed.) Encyclopedia of tribology, (2013) 2510-3.

[85]   Mughrabi, H: Fatigue, an everlasting materials problem - still en vogue. Procedia Engineering, 2, 1 (2010) 3–26.

[86]   Sohar, C: Lifetime controlling defects in tool steels. ISBN: 978-3-642-21645-9 (2011).

[87]   Kim, S; Hagiwara, M; Kawabe, Y; Kim, S: Internal crack initiation in high cycle fatigued Ti-15V-3Cr-3Al-3Sn alloys. Materials Science and Engineering: A, 334, 1-2 (2002) 73–8.

[88]   Hoefler A: Maschinenbau Physik: Zerstörende Werkstoffprüfung. [September 08, 2017]; Available from: http://www.ahoefler.de/maschinenbau/werkstoffkunde/zerstoerende-werkstoffpruefung/84-woehlerversuch.html.

[89]   Ritchie, R: Mechanisms of fatigue-crack propagation in ductile and brittle solids. International Journal of Fracture, 100, 1 (1999) 55–83.

[90]   Purushothaman, S; Tien, J: Generalized theory of fatigue crack propagation: Part I - derivation of thresholds. Materials Science and Engineering, 34, 3 (1978) 241–6.

[91]   Ellyin, F: Fatigue damage, crack growth and life prediction. ISBN: 978-9-401-07175-8 (1997).

[92]   Xiang, Y; Lu, Z; Liu, Y: Crack growth-based fatigue life prediction using an equivalent initial flaw model. Part I: Uniaxial loading. International Journal of Fatigue, 32, 2 (2010) 341–9.

[93]   Toyosada, M; Gotoh, K; Niwa, T: Fatigue life assessment for welded structures without initial defects: An algorithm for predicting fatigue crack growth from a sound site. International Journal of Fatigue, 26, 9 (2004) 993–1002.

[94]   Reuven, Y; Rubinstein, D; Kroese, D: Simulationa nd the Monte-Carlo method, ISBN: 978-0-470-17794-5 (2008).

[95]   Walpole, R; Myers, R; Myers, S; Ye, K: Probability & statistics for engineers and scientists. ISBN: 978-0-134-11585-6 (2016).

[96]   Burkart, K; Bomas, H; Zoch, H-W: Fatigue of notched case-hardened specimens of steel SAE 5120 in the VHCF regime and application of the weakest-link concept. International Journal of Fatigue, 33, 1 (2011) 59–68.

[97]   Bathe, K-J: Finite element procedures. ISBN: 978-0-979-00490-2 (2006).

[98]   Nocedal, J; Wright, S.: Numerical optimization. ISBN: 978-0-387-40065-5 (2006).

[99]   Gu, D; Hagedorn, Y; Meiners, W; Meng, G; Batista, R; Wissenbach, K; Poprawe, R: Densification behavior, microstructure evolution, and wear performance of selective laser melting processed commercially pure titanium. Acta Materialia, 60, 9 (2012) 3849–60.

[100]   Löber, L; Schimansky, F; Kühn, U; Pyczak, F; Eckert, J: Selective laser melting of a beta-solidifying TNM-B1 titanium aluminide alloy. Journal of Materials Processing Technology, 214, 9 (2014) 1852 60.

[101]   Attar, H; Calin, M; Zhang, L; Scudino, S; Eckert, J: Manufacture by selective laser melting and mechanical behavior of commercially pure titanium. Materials Science and Engineering: A, 593 (2014) 170–7.

[102]   Gu, D; Meiners, W; Wissenbach, K; Poprawe, R: Laser additive manufacturing of metallic components: Materials, processes and mechanisms. International Materials Reviews, 57, 3 (2012) 133–64.

[103]   Mumtaz, K; Hopkinson, N: Selective laser melting of thin wall parts using pulse shaping. Journal of Materials Processing Technology, 210, 2 (2010) 279–87.

[104]   Li, R; Liu, J; Shi, Y; Wang, L; Jiang, W: Balling behavior of stainless steel and nickel powder during selective laser melting process. The International Journal of Advanced Manufacturing Technology, 59, 9-12 (2012) 1025–35.

[105]   Kabir, M; Richter, H: Modeling of processing-induced pore morphology in an additively-manufactured Ti-6Al-4V alloy. Materials, 10, 2 (2017) 145.

[106]   Aboulkhair, N; Everitt, N; Ashcroft, I; Tuck, C: Reducing porosity in AlSi10Mg parts processed by selective laser melting. Additive Manufacturing, 1-4 (2014) 77–86.

[107]   Jerrard, P; Hao, L; Dadbakhsh, S; Evans, K: Consolidation behaviour and microstructure characteristics of pure aluminium and alloy powders following

selective laser melting processing, In: Hinduja, L (Ed.) Proceedings of the 36th International MATADOR Conference. London, (2010) 487–90.

[108]   Prashanth, K; Eckert, J: Formation of metastable cellular microstructures in selective laser melted alloys. Journal of Alloys and Compounds, 707 (2017) 27–34.

[109]   Dinda, G; Dasgupta, A; Bhattacharya, S; Natu, H; Dutta, B; Mazumder, J: Microstructural characterization of laser-deposited Al 4047 alloy. Metallurgical and Materials Transactions A, 44, 5 (2013) 2233–42.

[110]   Niendorf, T; Leuders, S; Riemer, A; Richard, H; Tröster, T; Schwarze, D: Highly anisotropic steel processed by selective laser melting. Metallurgical and Materials Transactions B, 44, 4 (2013) 794–6.

[111]   Niendorf, T; Leuders, S; Riemer, A; Brenne, F; Tröster, T; Richard, H; Schwarze, D: Functionally graded alloys obtained by additive manufacturing. Advanced Engineering Materials, 16, 7 (2014) 857–61.

[112]   Strano, G; Hao, L; Everson, R; Evans, K: Surface roughness analysis, modelling and prediction in selective laser melting. Journal of Materials Processing Technology, 213, 4 (2013) 589–97.

[113]   Pyka, G; Kerckhofs, G; Papantoniou, I; Speirs, M; Schrooten, J; Wevers, M: Surface roughness and morphology customization of additive manufactured open porous Ti6Al4V structures. Materials, 6, 10 (2013) 4737–57.

[114]   Bourell, D; Spierings, A; Herres, N; Levy, G: Influence of the particle size distribution on surface quality and mechanical properties in AM steel parts. Rapid Prototyping Journal, 17, 3 (2011) 195–202.

[115]   Gu, D; Shen, Y: Balling phenomena in direct laser sintering of stainless steel powder: Metallurgical mechanisms and control methods. Materials & Design, 30, 8 (2009) 2903–10.

[116]   Krol, M; Dobrzanski, L; Reimann, L; Czaja, I: Surface quality in selective laser melting of metal powders. Archives of Materials Science and Engineering, 60, 2 (2013) 87–92.

[117]   Guan, K; Wang, Z; Gao, M; Li, X; Zeng, X: Effects of processing parameters on tensile properties of selective laser melted 304 stainless steel. Materials & Design, 50 (2013) 581–6.

[118]   Hanzl, P; Zetek, M; Bakša, T; Kroupa, T: The influence of processing parameters on the mechanical properties of SLM parts. Procedia Engineering, 100 (2015) 1405–13.

[119]   Sufiiarov, V; Popovich, A; Borisov, E; Polozov, I: Selective laser melting of titanium alloy and manufacturing of gas-turbine engine part blanks. Tsvetnye Metally (2015) 76–80.

[120]   Shunmugavel, M; Polishetty, A; Littlefair, G: Microstructure and mechanical properties of wrought and additive manufactured Ti-6Al-4V cylindrical bars. Procedia Technology, 20 (2015) 231–6.

[121]   Buchbinder, D; Schleifenbaum, H; Heidrich, S; Meiners, W; Bültmann, J: High
        power selective laser melting (HP SLM) of aluminum parts. Physics Procedia, 12
        (2011) 271–8.

[122]   Manfredi, D; Calignano, F; Krishnan, M; Canali, R; Ambrosio, E; Atzeni, E:
        From powders to dense metal parts: Characterization of a commercial AlSiMg
        alloy processed through direct metal laser sintering. Materials, 6, 3 (2013) 856–
        69.

[123]   Suryawanshi, J; Prashanth, K; Scudino, S; Eckert, J; Prakash, O; Ramamurty, U:
        Simultaneous enhancements of strength and toughness in an Al-12Si alloy
        synthesized using selective laser melting. Acta Materialia, 115 (2016) 285 94.

[124]   Delgado, J; Ciurana, J; Rodríguez, C: Influence of process parameters on part
        quality and mechanical properties for DMLS and SLM with iron-based materials.
        The International Journal of Advanced Manufacturing Technology, 60, 5-8
        (2012) 601–10.

[125]   Edwards, P; Ramulu, M: Fatigue performance evaluation of selective laser melted
        Ti–6Al–4V. Materials Science and Engineering: A, 598 (2014) 327–37.

[126]   Wycisk, E; Kranz, J; Emmelmann, C: Fatigue strength of light weight structures
        produced by laser additive manufacturing in TiAl6V4, In: Tavares, S (Ed.) 1st
        international conference of the international journal of structural integrity 25-28
        June (2012).

[127]   Spierings, A; Starr, T; Wegener, K: Fatigue performance of additive
        manufactured metallic parts. Rapid Prototyping Journal, 19, 2 (2013) 88–94.

[128]   Rafi, H; Starr, T; Stucker, B: A comparison of the tensile, fatigue, and fracture
        behavior of Ti–6Al–4V and 15-5 PH stainless steel parts made by selective laser
        melting. The International Journal of Advanced Manufacturing Technology, 69,
        5-8 (2013) 1299–309.

[129]   Siddique, S; Imran, M; Rauer, M; Kaloudis, M; Wycisk, E; Emmelmann, C;
        Walther, F: Computed tomography for characterization of fatigue performance of
        selective laser melted parts. Materials & Design, 83 (2015) 661–9.

[130]   Sehrt, J; Witt, G: Dynamic strength and fracture toughness analysis of beam
        melted parts, In: Hinduja, S; Lin, L (Eds.) Proceedings of the 36th international
        MATADOR conference. ISBN: 978-1-849-96431-9 (2010) 385-88.

[131]   Wycisk, E; Siddique, S; Herzog, D; Walther, F; Emmelmann, C: Fatigue
        performance of laser additive manufactured Ti–6Al–4V in very high cycle fatigue
        regime up to 1E9 cycles. Frontiers in Materials, 2, 72 (2015) 1–8.

[132]   Leuders, S; Thoene, M; Riemer, A; Niendorf, T; Tröster, T; Richard, H; Maier,
        H: On the mechanical behaviour of titanium alloy TiAl6V4 manufactured by
        selective laser melting: Fatigue resistance and crack growth performance.
        International Journal of Fatigue, 48 (2013) 300–7.

[133]   Riemer, A; Leuders, S; Thoene, M; Richard, H; Tröster, T; Niendorf, T: On the
        fatigue crack growth behavior in 316L stainless steel manufactured by selective
        laser melting. Engineering Fracture Mechanics, 120 (2014) 15–25.

[134]   Wang, F: Mechanical property study on rapid additive layer manufacture Hastelloy® X alloy by selective laser melting technology. The International Journal of Advanced Manufacturing Technology, 58, 5 (2012) 545–51.

[135]   Maskery, I; Aboulkhair, N; Tuck, C; Wildman, R; Ashcroft, I; Everitt, N; Hague, R: Fatigue performance enhancement of selectively laser melted aluminium alloy by heat treatment. In: 26th annual international solid freeform fabrication symposium (2015).

[136]   Okyar, A; Uzunsoy, D; Ozsoy, B: Comparison of fatigue crack growth rate of selective laser sintered rapid steel via computational fracture mechanics. International Journal of Materials Research, 105, 6 (2014) 552–6.

[137]   Cain, V; Thijs, L; Humbeeck, J; Hooreweder, B; Knutsen, R: Crack propagation and fracture toughness of Ti6Al4V alloy produced by selective laser melting. Additive Manufacturing, 5 (2015) 68–76.

[138]   Edwards, P; Ramulu, M: Effect of build direction on the fracture toughness and fatigue crack growth in selective laser melted Ti-6Al-4 V. Fatigue and Fracture of Engineering Materials & Structures, 38, 10 (2015) 1228–36.

[139]   Stanzl-Tschegg, S: Very high cycle fatigue measuring techniques. International Journal of Fatigue, 60 (2014) 2–17.

[140]   Bathias, C: Piezoelectric fatigue testing machines. International Journal of Fatigue, 28, 11 (2006) 1438–45.

[141]   Warmuzek, M: Aluminum-silicon casting alloys: An atlas of microfractographs. Materials Park: ASM International; (2004).

[142]   Davis, J: Corrosion of aluminum and aluminum alloys. ISBN: 978-0-87170-629-4 (1999).

[143]   Ahmad Z: (Ed.) Aluminium alloys - New trends in fabrication and applications; ISBN: 978-9-535-10861-0 (2012).

[144]   Murray, J; McAlister, A: The Al-Si (aluminum-silicon) system. Bulletin of Alloy Phase Diagrams, 5, 1 (1984) 74–84.

[145]   Shankar, S; Riddle, Y; Makhlouf, A: Eutectic solidification of aluminum-silicon alloys. Metallurgical and Materials Transactions A, 35, A (2004) 3038–43.

[146]   Torobian, H; Patahak, J; Tiwari, S: Wear characteristics of Al-Si alloys. Wear, 172, 1 (1994) 49–58.

[147]   Miller, W; Zhuang, L; Bottema, J; Wittebrood, A; Smet, P; Haszler, A; Vieregge, A: Recent development in aluminium alloys for the automotive industry. Materials Science and Engineering: A, 280, 1 (2000) 37–49.

[148]   Siddique, S; Tenkamp, J; Walther, F: Influence of process-induced defects on the high cycle and very high cycle fatigue behavior of additively-manufactured AlSi12 structures, In: Richard, H (Ed.) 1. Tagung des DVM-AK Additiv gefertigte Bauteile und Strukturen. ISSN: 2509-8772 (2016) 83-90.

[149]   Siddique, S; Awd, M; Walther, F: Influence of hybridization by selective laser melting on the very high cycle fatigue behaviour of aluminium alloys, In: Zimmermann, M; Christ, H-J (Eds.) Proceedings of the 7th international conference on very high cycle fatigue. ISBN: 978-3-000-56960-9 (2017) 229-34.

[150]   Siddique, S; Tenkamp, J; Walther, F: Einfluss prozessinduzierter Defekte auf das Ermüdungsverhalten additiv-gefertigter AlSi12-Strukturen bei hohen und sehr hohen Lastspielzahlen, In: Richard, H; Schramm, B; Zipsner, T (Eds.) Additive Fertigung von Bauteilen und Strukturen. ISBN: 978-3-658-17779-9 (2017) 215-26.

[151]   Siddique, S; Walther, F: Selective laser melting: Mechanical performance of light-weight alloys, In: Thornton, A. (Ed.) Additive manufacturing (AM): Emerging technologies, applications and economic implications. ISBN: 978-1-634-63850-0 (2015) 75-109.

[152]   Siddique, S; Walther, F: Fatigue and fracture reliability of additively manufactured Al-4047 and Ti-6Al-4V alloys for automotive and space applications, In: Bajpai, RP; Chandrasekhar, U (Eds.) Innovative design and development practices in aerospace and automotive engineering. ISBN: 978-9-811-01770-4 (2016) 19-25.

[153]   Siddique, S; Imran, M; Walther, F: Very high cycle fatigue and fatigue crack propagation behavior of selective laser melted AlSi12 alloy. International Journal of Fatigue, 94, 2 (2017) 246–54.

[154]   Siddique, S; Imran, M; Wycisk, E; Emmelmann, C; Walther, F: Fatigue assessment of laser additive manufactured AlSi12 eutectic alloy in the very high cycle fatigue (VHCF) range up to 1E9 cycles. Materials Today: Proceedings, 3 (2016) 2853–60.

[155]   Siddique, S; Awd, M; Tenkamp, J; Walther, F: Development of a stochastic approach for fatigue life prediction of AlSi12 processed by selective laser melting. Engineering Failure Analysis, 79 (2017) 34–50.

[156]   Walther, F; Eifler, D: Short-time procedure for the determination of Woehler and fatigue life curves using mechanical, thermal and electrical data. Journal of Solid Mechanics and Materials Engineering, 2, 4 (2008) 507–18.

[157]   Walther, F; Eifler, D: PHYBAL - Kurzzeitverfahren zur Berechnung der Lebensdauer metallischer Werkstoffe auf der Basis physikalischer Messgrößen. Materials Testing, 50, 3 (2008) 142–9.

[158]   Murakami, Y: Metal fatigue: Effects of small defects and nonmetallic inclusions. ISBN: 978-0-080-44064-4 (2002).

[159]   Siddique, S; Awd, M; Tenkamp, J; Walther, F: High and very high cycle fatigue failure mechanisms in selective laser melted aluminium alloys. Journal of Materials Research (2017) in press.

[160]   Asghar, Z; Requena, G; Kubel, F: The role of Ni and Fe aluminides on the elevated temperature strength of an AlSi12 alloy. Materials Science and Engineering: A, 527, 21-22 (2010) 5691–8.

[161]   Simchi, A: Direct laser sintering of metal powders: Mechanism, kinetics and microstructural features. Materials Science and Engineering: A, 428, 1-2 (2006) 148–58.

[162]   Monroy, K; Delgado, J; Ciurana, J: Study of the pore formation on CoCrMo alloys by selective laser melting manufacturing process. Procedia Engineering, 63 (2013) 361–9.

[163]   DIN: Aluminium and aluminium alloys - Castings - Chemical composition and mechanical properties. EN 1706:2013. Berlin: Beuth (2013).

[164]   Li, X; Li, J; Ding, W; Zhao, S; Chen, J: Stress relaxation in tensile deformation of 304 stainless steel. Journal of Materials Engineering and Performance, 26, 2 (2017) 630–5.

[165]   Nakai, Y; Kusukawa, T; Maeda, K: Grain size effect on fatigue crack initiation condition observed by using atomic-force microscopy, In: 10th international conference on fracture, Hawaii, (2001).

[166]   Martienssen, W; Warlimont, H: Springer handbook of condensed matter and materials data. ISBN: 978-3-540-30437-1 (2005).

[167]   Bruckner, A; Tschegg, E; Schuller, A: Mechanical properties of electron-beam-melted AlSi12 surface layers. Journal of Materials Science, 25 (1990) 5220–4.

[168]   Mower, T; Long, M: Mechanical behavior of additive manufactured, poweder-bed laser-fused materials. Materials Science and Engineering: A, 651 (2016) 198–213.

[169]   Mayer, H: Recent developments in ultrasonic fatigue. Fatigue & Fracture of Engineering Materials and Structures, 39, 1 (2016) 3–29.

[170]   Guennec, B; Ueno, A; Sakai, T; Takanashi, M; Itabashi, Y: Effect of loading frequency in fatigue properties and micro-plasticity behavior of JIS S15C low carbon steel, In: 13th international conference on fracture, Beijing, (2013).

[171]   Tsutsumi, N; Murakami, Y; Doquet, V: Effect of test frequency on fatigue strength of low carbon steel. Fatigue and Fracture of Engineering Materials & Structures, 32, 6 (2009) 473–83.

[172]   Furuya, Y; Torizuka, S; Takeuchi, E; Bacher-Hoechst, M; Kuntz, M: Ultrasonic fatigue testing on notched and smooth specimens of ultrafine-grained steel. Materials & Design, 37 (2012) 515–20.

[173]   Furuya, Y; Matsuoka, S; Abe, T; Yamaguchi, K: Gigacycle fatigue properties for high-strength low-alloy steel at 100 Hz, 600 Hz, and 20 kHz. Scripta Materialia, 46, 2 (2002) 157–62.

[174]   Mayer, H; Schuller, R; Fitzka, M: Fatigue of 2024-T351 aluminium alloy at different load ratios up to $10^{\wedge}10$ cycles. International Journal of Fatigue, 57 (2013) 113–9.

[175]   Mayer, H; Papakyriacou, M; Pippan, R; Stanzl-Tschegg, S: Influence of loading frequency on the high cycle fatigue properties of AlZnMgCu1.5 aluminium alloy. Materials Science and Engineering: A, 314, 1-2 (2001) 48–54.

[176]   Stanzl-Tschegg, S; Mayer, H; Schuller, R; Przeorski, T; Krug, P: Fatigue properties of spray formed hypereutectic aluminium silicon alloy DISPAL® S232 at high and very high numbers of cycles. Materials Science and Engineering: A, 538 (2012) 327–34.

[177]   Zhu, X; Jones, J; Allison, J: Effect of frequency, environment, and temperature on fatigue behavior of E319 cast aluminum alloy: Stress-controlled fatigue life response. Metallurgical and Materials Transactions A, 39, 11 (2008) 2681–8.

[178]   Siddique, S; Wycisk, E; Tenkamp, J; Hoops, K; Behrens, G; Emmelmann, C; Walther, F: Mechanical performance of hybrid aluminum structures manufactured by combination of laser additive manufcaturing and conventional machining processes, In: Borsutzki, M; Moninger, G (Eds.) Werkstoffprüfung 2015 - Fortschritte in der Werkstoffprüfung für Forschung und Praxis. ISBN: 978-3-514-00816-8 (2015) 157-62.

[179]   Liu, J: Advanced aluminium and hybrid aero structures for future aircraft. Materials Science Forum, 519-521 (2006) 1233–8.

[180]   Klocke, F; Roderburg, A; Zeppenfeld, C: Design methodology for hybrid production processes. Procedia Engineering, 9 (2011) 417–30.

[181]   Awd, M: Statistical numerical investigation of fatigue behavior of selective laser melted aluminum alloys [Master thesis]. Dortmund: Technical University Dortmund; (2017).

[182]   Leuders, S; Vollmer, M; Brenne, F; Tröster, T; Niendorf, T: Fatigue strength prediction for titanium alloy TiAl6V4 manufactured by selective laser melting. Metallurgical and Materials Transactions A, 46 (2015) 3816-23.

The following student works were supervised in the framework of this dissertation.

- Awd, M: Development of a stochastic approach for fatigue life prediction of AlSi12 alloy processed by selective laser melting. Project work (2016).

- Awd, M: Statistical numerical investigation of fatigue behavior of selective laser melted aluminum alloys. Master thesis (2017).

The following articles were pre-published in the framework of this dissertation.

- Wycisk, E; Emmelmann, C; Siddique, S; Walther, F: High cycle fatigue (HCF) performance of Ti-6Al-4V alloy processed by selective laser melting. Advanced Materials Research, 816-817 (2013) 134-9.

- Wycisk, E; Solbach, A; Siddique, S; Herzog, D; Walther, F; Emmelmann, C: Effects of defects in laser additive manufactured Ti-6Al-4V on fatigue properties. Physics Procedia, 56 (2014) 371-8.

- Wycisk, E; Siddique, S; Herzog, D; Walther, F; Emmelmann, C: Fatigue performance of laser additive manufactured Ti-6Al-4V in very high cycle fatigue regime up to 10^9 cycles. Frontiers in Materials, 2:72 (2015) 1-8.

- Siddique, S; Wycisk, E; Frieling, G; Emmelmann, C; Walther, F: Microstructural and mechanical properties of selective laser melted Al 4047. Applied Mechanics and Materials, 752-753 (2015) 485-90.

■   Siddique, S; Imran, M; Wycisk, E; Emmelmann, C; Walther, F: Influence of process-induced microstructure and imperfections on mechanical properties of AlSi12 processed by selective laser melting. Journal of Materials Processing Technology, 221 (2015) 205-13.

■   Siddique, S; Imran, M; Rauer, M; Kaloudis, M; Wycisk, E; Emmelmann, C; Walther, F: Computed tomography for characterization of fatigue performance of selective laser melted parts. Materials & Design, 83 (2015) 661-9.

■   Siddique, S; Walther, F: Selective laser melting: Mechanical performance of light-weight alloys. In: Thornton, A (Ed.) Additive manufacturing (AM): Emerging technologies, applications and economic implications, ISBN: 978-1-63463-850-0 (2015) 75-109.

■   Siddique, S; Walther, F: Fatigue and fracture reliability of additively manufactured Al-4047 and Ti-6Al-4V alloys for automotive and aerospace applications. In: Bajpai, RP; Chandrasekhar, U (Eds.) Innovative design and development practices in aerospace and automotive engineering, ISBN: 978-9-811-01770-4 (2016) 19-25.

■   Siddique, S; Wycisk, E; Tenkamp, J; Hoops, K; Behrens, G; Emmelmann, C; Walther, F: Mechanical performance of hybrid aluminum structures manufactured by combination of laser additive manufacturing and conventional machining processes. In: Borsutzki, M; Moninger, G (Eds.) Werkstoffprüfung 2015 – Fortschritte in der Werkstoffprüfung für Forschung und Praxis, ISBN 978-3-514-00816-8 (2015) 157-62.

■   Siddique, S; Tenkamp, J; Walther, F: Einfluss prozessinduzierter Defekte auf das Ermüdungsverhalten additiv-gefertigter AlSi12-Strukturen bei hohen und sehr hohen Lastspielzahlen. In: Richard, H-A (Ed.) 1. Tagung des DVM-AK Additiv gefertigte Bauteile und Strukturen, ISSN: 2509-8772 (2016) 83-90.

■   Siddique, S; Imran, M; Wycisk, E; Emmelmann, C; Walther, F: Fatigue assessment of laser additive manufactured AlSi12 eutectic alloy in the very high cycle fatigue (VHCF) range up to 1E9 cycles. Materials Today: Proceedings, 3 (2016) 2853–60.

■   Siddique, S; Imran, M; Walther, F: Very high cycle fatigue and fatigue crack propagation behavior of selective laser melted AlSi12 alloy. International Journal of Fatigue, 94, 2 (2017) 246-54.

■   Siddique, S; Awd, M; Tenkamp, J; Walther, F: Development of a stochastic approach for fatigue life prediction of AlSi12 alloy processed by selective laser melting. Engineering Failure Analysis, 79 (2017) 34-50.

■   Siddique, S; Tenkamp, J; Walther, F: Einfluss prozessinduzierter Defekte auf das Ermüdungsverhalten additiv-gefertigter AlSi12-Strukturen bei hohen und sehr hohen Lastspielzahlen. In: Richard, H; Schramm, B; Zipsner, T (Eds.)

Additive Fertigung von Bauteilen und Strukturen, ISBN: 978-3-658-17779-9 (2017) 215-26.

■   Tenkamp, J; Siddique, S; Walther, F: Detaillierte Untersuchungen zur Bewertung des Ermüdungsverhaltens von Bauteilen bei hohen Lastspielzahlen. Qualität und Zuverlässigkeit, 5 (2017) 88-90.

■   Siddique, S; Awd, M; Walther, F: Influence of hybridization by selective laser melting on the very high cycle fatigue behaviour of aluminium alloys. In: Zimmermann, M; Christ, H-J (Eds.) VHCF7- Proc. of the 7th international conference on very high cycle fatigue, ISBN: 978-3-00-056960-9 (2017) 229-34.

■   Siddique, S; Awd, M; Tenkamp, J; Walther, F: High and very high cycle fatigue failure mechanisms in selective laser melted aluminium alloys. Journal of Materials Research, 32, 23 (2017) 4296-4304.

Curriculum Vitae

Personal

Name:	Shafaqat Siddique
Date of birth:	22.12.1982 in Kasur, Pakistan
Marital status:	Married

Education

1988-1998	Matriculation, Kasur
1998-2000	Intermediate Pre-Engineering, GC Lahore
2001-2005	B.Sc. Mechanical Engineering, UET Lahore
2005-2008	M.Sc. Manufacturing Engineering, UET Lahore
2009-2012	M.Sc. International Production Management, TU Hamburg

Professional carrier

2005-2006	Lecturer, University of Central Punjab, Lahore
2007-2009	Lecturer, The University of Lahore, Lahore
Since 2013	Scientific Assistant, Department of Materials Test Engineering, TU Dortmund

© Springer Fachmedien Wiesbaden GmbH, part of Springer Nature 2019
S. Siddique, *Reliability of Selective Laser Melted AlSi12 Alloy for Quasistatic and Fatigue Applications*, Werkstofftechnische Berichte | Reports of Materials Science and Engineering, https://doi.org/10.1007/978-3-658-23425-6